This book is
dedicated to
Margaret Bishop -
a friend and
geologist ahead
of her time.

QUICK LOOK TECHNIQUES
FOR
PROSPECT EVALUATION

DANIEL J. TEARPOCK
RICHARD E. BISCHKE
AND
JOSEPH L. BREWTON

Subsurface Consultants & Associates, Inc., Lafayette, Louisiana 70508

Tearpock, Daniel J.
   Quick look techniques for prospect evaluation / Daniel J.
Tearpock, Richard E. Bischke and Joseph L. Brewton.
   p. cm.
   Includes bibliographical references and index.
   ISBN 0-9642961-0-1

   1. Petroleum--Geology. 2. Geological mapping. 3. Petroleum--
Prospecting.   I. Title.

TN870.5.T43 1994   553.2'8
                                    QBI94-1708

Library of Congress Catalog Card Number: 94-68541

Editorial/production supervision: Reneé H. Ory
Interior design: Nicole McMorris
Printing: Touché Printing
Cover design: Graphics Plus

The publisher offers discounts on this book when ordered
in bulk quantities. For more information, write:
       Special Sales/College Marketing
       Subsurface Consultants & Associates, LLC
       400 E. Kaliste Saloom Road, Suite 2000
       Lafayette, LA 70508

Printed in the United States of America
10 9 8 7 6 5 4 3 2

ISBN 0-9642961-0-1

TABLE OF
CONTENTS

PREFACE   xi

ACKNOWLEDGEMENTS   xii

**CHAPTER 1 PHILOSOPHICAL DOCTRINE OF
            SUBSURFACE PROSPECT MAPPING   1**

INTRODUCTION   1

SUBSURFACE MAPPING - PHILOSOPHICAL DOCTRINE   2

CHAPTER 2   DRY HOLE ANALYSIS   6

INTRODUCTION   6

FOUR BASIC GEOLOGIC REASONS FOR DRY HOLES   7

WHAT! ANOTHER DRY HOLE?   11

CHAPTER 3   **GENERAL STRUCTURE MAP QUICK LOOK TECHNIQUES**   **12**

INTRODUCTION   12

EQUAL-SPACED CONTOURING METHOD   12

CONTOUR COMPATIBILITY BETWEEN CLOSELY SPACED HORIZONS   16

POROSITY TOP VERSUS STRUCTURE TOP   19

RESTORED TOPS AND THEIR USE   23

THE MAP JUST DOESN'T LOOK RIGHT   28

HIGH-LOW'S   28

CHAPTER 4   **FAULTED STRUCTURE MAP QUICK LOOK TECHNIQUES**   **33**

INTRODUCTION   33

FAULT SURFACE ANALYSIS   34

     Normal Faults   34
     Thrust and Reverse Faults   37

STRUCTURAL COMPATIBILITY ACROSS FAULTS   38

INTEGRATED FAULT MAP/STRUCTURE MAP   40

HONORING FAULT DATA FROM WELL LOGS IN THE
CONSTRUCTION OF COMPLETED STRUCTURE MAPS   46

    Normal Faults   46

REVERSE FAULT/STRUCTURE INTEGRATION   54

ADDITIVE PROPERTY OF FAULTS   57

IMPLIED FAULT ANALYSIS ON STRUCTURE MAPS   60

THE "RULE OF 45"   65

PITFALLS OF FAULT GAP WIDTH   67

SCREW FAULTS   70

SAUCER FAULTS   74

ODD NUMBER OF CONTOURS TERMINATING AROUND
A FAULT   76

CHAPTER 5   SEISMIC
                 QUICK LOOK TECHNIQUES   78

INTRODUCTION   78

DATA QUALITY AND VALIDITY   78

CORRELATION AND MIS-TIE TECHNIQUES   87

THREE-DIMENSIONAL STRUCTURAL GEOMETRY   95

QUICK LOOK MAPPING   100

CHAPTER 6   COMPRESSIONAL STRUCTURAL GEOLOGY QUICK LOOK TECHNIQUES   102

INTRODUCTION   102

A QUICK LOOK AT BED LENGTHS   103

A QUICK LOOK AT BED THICKNESS   104

RETRODEFORMATION   109

THIN SKINNED TECTONICS   111

FOLD SHAPE IS RELATED TO FAULT SHAPE   113

BALANCING AND THRUST FAULT STEP-UP
ANGLES   114

DO THE AXIAL SURFACES BISECT THE FOLD
LIMBS - OR OFF STRUCTURE AGAIN?   117

    Proper Well Position   120

THRUST FAULT POSITION   125

    Dip Domain Analysis:
    Rapidly Locating Thrusts And Plays   126
    Quickly Recognizing Décollements   130
    Rapidly Locating Additional Prospects   130

KINK OR A FAULT?   131

FAULTS OR A BOX FOLD?   134

LIMITED STRUCTURAL STYLES   137

MAP AND CROSS-SECTION CONSISTENCY   138

**DUPLEX STRUCTURES: HOW TO
RECOGNIZE MULTIPLE PLAYS**   141

GROWTH   146

    Rapid Sedimentation Rates   153
    Slow Sedimentation Rates   156

**QUICKLY EVALUATING "BALANCED"
CROSS-SECTIONS**   157

    Balanced Section Pitfalls   157
        Passive Roof Duplex   159
        Fish Hook Faults   159
        Multi-Play Pitfall   161

FAULT PROPAGATION FOLD DIP ANALYSIS   162

CHAPTER 7   **EXTENSIONAL STRUCTURAL
GEOLOGY
QUICK LOOK TECHNIQUES**   **168**

INTRODUCTION   168

**LISTRIC FAULT SHAPE DETERMINES
ROLLOVER SHAPE**   168

viii

RAPIDLY LOCATING MAJOR ROLLOVER HIGHS   169

ANTILISTRIC FAULT BENDS: A SAND INDICATOR   178

AXIAL SURFACES: LOCATING FAULT POSITION   183

RELATIONSHIP BETWEEN NORMAL FAULT SHAPE AND PROSPECT SHAPE AND POSITION   186

COMMON FOLD PATTERNS   188

Rolled Up Beds at the Front of a Rollover   189
Half-Graben or Monoclinal Rollover
Structure   192
Fanning of Beds Dips   195
Basinward Closure   195
Review of Methods and Generalizations   196
Case History   198

GROWTH QLTS   201

Using Antilistric Fault Terminations
to Determine Growth   203
Using Axial Surfaces to Determine Growth   204
Slope of Axial Surface as a Potential
Hydrocarbon Indicator   209

EXPANDED REFLECTIONS: HOW TO IDENTIFY DEEPER GROWTH ZONES   211

EXPANSION INDICES AND $\Delta d/d$ DIAGRAMS (A 30 Minute Technique)   212

DOWNWARD DYING GROWTH FAULT PITFALL   220

Strike Ramp Pitfall   222

CHAPTER 8   DIRECTIONAL WELL
QUICK LOOK TECHNIQUES   224

INTRODUCTION   224

VERTICAL SEPARATION AND DIRECTIONALLY
DRILLED WELLS   224

PRODUCTION FAULTS OR WATER LEVEL
CHANGES-REAL OR THE RESULT OF
DIRECTIONAL SURVEY UNCERTAINTIES   229

DIRECTIONALLY DRILLED WELLS, NET SAND,
AND NET PAY ISOCHORE MAPS   232

CHAPTER 9   ISOCHORE MAPS
QUICK LOOK TECHNIQUES   237

INTRODUCTION   237

THE WHARTON METHOD FOR NET PAY ISOCHORE
MAPS   237

TOO MUCH NET PAY   241

ISOCHORE VERSUS ISOPACH   246

CHAPTER 10   PROJECT EVALUATIONS   248

x

INTRODUCTION   248

QUICK LOOK TECHNIQUE PROSPECT REVIEW QUESTIONS   248

QLT QUESTIONS   249

PROJECTS   251

PROJECT ONE   252

PROJECT TWO   254

PROJECT THREE   256

PROJECT FOUR   258

PROJECT FIVE   260

PROJECT MAP SIX   262

PROJECT MAP SEVEN   264

PROJECT MAP EIGHT   266

PROJECT MAP NINE   268

PROJECT MAP TEN   270

PROJECT MAP ELEVEN   272

PROJECT MAP TWELVE   275

REFERENCES   277

INDEX   283

PREFACE

Many people are involved with either preparing or reviewing and evaluating prospect maps. Significant investment decisions are often made based on prospects presented in the form of completed structure maps. This text is written for those who need a rapid method for reviewing geologic maps. Those interested in the *detailed preparation* of subsurface maps and profiles should consult Tearpock and Bischke (1991).

Too often, geologic maps and cross sections are prepared without giving adequate consideration to the three-dimensional validity of the interpretation. Investment decisions are frequently made based on maps, cross sections, and seismic sections that have not undergone a detailed technical review.

Prospect maps are made to sell an idea. Prospectors must be optimistic in their work, or they may be incapable of identifying hydrocarbon potential. When potential is identified, a prospector must be able to convince someone to drill or invest in the prospect. Prospect interpretations may be optimistic, pessimistic, unrealistic, or downright impossible. How does a company or investor distinguish a great prospect from a mediocre one or a reasonable prospect interpretation from an impossible one? This recognition should be accomplished by applying the same expertise to evaluate prospects that is required to generate them.

Detailed evaluations must often be conducted in a limited amount of time. This can only be done if the evaluator has sound geologic expertise and understands the use of correct interpretive and mapping techniques. With experience, the quality of prospect maps and sections (seismic or geologic) can often be quickly evaluated by using a number of **Quick Look Techniques (QLTs)**.

This text is a companion to <u>Applied</u> <u>Subsurface</u> <u>Geological</u> <u>Mapping</u> by Tearpock and Bischke (1991). The QLTs outlined in this book provide a rapid means of evaluating prospects. However, the use of the QLTs relies heavily on the evaluator's understanding of structural geology and a variety of subsurface mapping and cross section techniques. Any specific QLT should not be applied without a thorough knowledge of the actual mapping technique or techniques from which the QLT was derived. The application of a QLT without knowledge of the actual technique being applied might lead to incorrect assumptions or decisions that are not accurate.

Quick Look Techniques are powerful tools when applied correctly. Their use can provide accurate and rapid answers about the quality of a prospect or geologic interpretation. When decisions are critical in selecting prospects for investment dollars, Quick Look Techniques are powerful tools.

ACKNOWLEDGEMENTS

REVIEWERS

We would like to thank several geoscientists who reviewed all or portions of the text. Peter Verrell (Chevron - retired), Jim Clement (Shell - retired), and Allan Nunns (Chevron) reviewed the entire text and made many helpful additions, improvements, and suggestions. Ron Hartman (International Exploration - retired), John Shaw (Texaco), Brian Lock, and Cathy Bishop, also made significant improvements and additions to portions of the text.

CONTRIBUTORS

We thank those persons, companies, and societies that contributed certain figures which have improved the quality of the textbook. American Association of Petroleum Geologists; American Journal of Earth Science; Chris Banks; Muzium Brunei; Coalinga Corporation; Denver Geological Society; M.C. Escher/Cordon Art; European Association of Exploration Geophysicists; Exxon USA, Company; Gulf Coast Association of Geological Societies; JEBCO Seismic Inc., Houston, Texas; Prentice-Hall, Inc.; Journal of Petroleum Technology; Journal of Structural Geology; National Council of Canada; Nippon Western US Company, LTD; Philippine National Oil Company; John Ramsey; Rocky Mountain Association of Geologists; Royal Society of London; Seitel Data Corporation; Donald Stone; John Suppe; TGS/GECO; Tenneco Oil Company; Texaco USA; Paul Tucker; U.S. Geological Survey; John Warburton; Hongbin Xiao; and Howard Yorston.

SUPPORT PERSONNEL

Support and management personnel are also critical to completing a text. Reneé Ory conducted the general text management and arranged the printing, grammatical reviews, the permissions and form approvals. Nicole McMorris typed and formatted the text, and organized the figures. Sona Dombourian conducted the grammatical review. Elsie Bischke and Darlene Hebert helped type several chapters. Cathy Bishop and Susan Deering contributed organizational support during the preparation of the textbook.

DRAFTING

The drafting was conducted by two professionals, Carmen Speer and Steve Nelson (independent). Danielle Hitt, Gwen Faulk, Lynette Bouton, and Nadine Biessenberger collated the text.

SPECIAL THANKS

We give special thanks to the many people in numerous oil companies that have submitted examples, made suggestions, and provided support and interest. Tenneco Oil Company provided early support on the subject of Quick Look Techniques. Jim Harris worked with Dan Tearpock at Tenneco Oil Company to establish the methodology for using Quick Look Techniques and their importance in evaluating prospects. Jim's early work was significant and we give him a special thanks.

For several years Mr. Greg Jones and Exxon USA have supported the concept and importance of Quick Look Techniques. Mr. Jones has provided many suggestions that have improved the overall content of the text.

CHAPTER ONE

PHILOSOPHICAL DOCTRINE
OF
SUBSURFACE PROSPECT MAPPING

INTRODUCTION

A reasonable and accurate subsurface interpretation and the generation of a hydrocarbon prospect must begin with an appropriate interpretation philosophy which incorporates a number of important and essential steps that should be followed in the generation of a prospect. The purpose of generating a prospect is to find and develop previously undiscovered hydrocarbons. There are many factors that underlie successful prospects; however, the factors that are probably the most important are: (1) each geologic interpretation must have **three-dimensional geologic and geometric validity,** (2) all the appropriate available data must be incorporated into the interpretation, and (3) correct mapping techniques must be used.

We, as prospectors, primarily deal with one-dimensional (well logs) and two-dimensional (conventional seismic sections) data which are used to develop a three-dimensional interpretation. One and two-dimensional data can be **misleading** or **misunderstood** if they are not interpreted correctly and integrated into a three-dimensional interpretation. Many incorrect geologic interpretations result from the use of isolated subsurface data (Tearpock and Bischke 1991). The job of a prospector is difficult enough when all the data are used. It is difficult, if not impossible, to prepare

1

accurate interpretations if only some of the data are used. What data do you ignore? How does it impact the interpretation?

SUBSURFACE MAPPING-PHILOSOPHICAL DOCTRINE

The philosophical doctrine we present here can serve as a *guide* for someone involved in generating subsurface prospects and as a *check list* for those who are evaluating them. If you are involved in reviewing and evaluating prospects, you may be in a position of responsibility, thus deciding where your company is going to place its investment dollars. Each company investing in prospect drilling expects positive economic results from its exploration efforts. The following philosophy can be used to help you make better investment decisions in oil and gas prospecting.

1. **All subsurface interpretations must be geologically and geometrically valid in three-dimensions.**

2. An interpreter must have a **sound background** in structural geology, stratigraphy, sedimentology, and other related disciplines for the *tectonic setting* being worked.

3. All subsurface data **must** be used to develop a reasonable and accurate subsurface geologic interpretation.

4. **All important and relevant geologic surfaces must be mapped and integrated** to arrive at a reasonable and accurate subsurface picture. These include surfaces such as formations, faults, unconformities, and salt.

5. The mapping of multiple horizons **is essential** to develop reasonably correct, three-dimensional interpretations of complexly deformed areas.

6. **Accurate correlations** (well log and seismic) are required for reliable geologic interpretations.

7. The use of correct mapping techniques and methods is essential to generate reasonable and accurate subsurface interpretations.

8. **Interpretive contouring** is the most acceptable method of contouring subsurface features such as faults, salt, unconformities, and structures.

9. All work should be **documented**.

10. **Sufficient time** and detail is required to generate reliable interpretations and prospects.

1. All subsurface interpretations must be geologically and geometrically valid in three dimensions. Subsurface data are either one-dimensional (well log) or two-dimensional (cross sections and conventional seismic sections); however, these data are used to generate a three-dimensional picture. Even though it is intuitive that all interpretations must be valid in three-dimensions, too often subsurface structure maps, cross sections, and seismic interpretations are made without much consideration given to establishing a three-dimensional framework or verifying that the interpretation *is even possible* in three dimensions.

2. An interpreter must have a sound background in structural geology, stratigraphy, sedimentology and other related disciplines for the tectonic setting being worked. When an interpretation is made in a particular tectonic setting, the interpreter **must** know as much about the structural geology of the area as possible, so that the interpretation represents geology that is known to fit the structural style of the area. In frontier areas, an exploration group must take it upon itself to develop a geologic framework with a regional prospective. This will insure that a regionally balanced, big-picture, with structural and stratigraphic models have been developed. A limited understanding of structural geology is one of the shortcomings in numerous geologic interpretations that results in unrealistic or even impossible interpretations in three dimensions. This subject cannot be covered here in any detail, but it is extremely important.

3. All subsurface data **must** be used to develop a reasonable and accurate subsurface interpretation. Each of us has a formidable task in analyzing structures that are thousands of feet below the surface. The data available are limited to begin with and are typically one and two-dimensional. A body of data itself can be confusing with respect to true subsurface relationships. For example, cross sections and seismic sections can misrepresent true three-dimensional subsurface relationships by the simple nature of their orientations. All data, (well log, seismic, production, paleo, etc.) must be integrated into an interpretation if it is to be considered sound and viable.

4. All important and relevant geologic surfaces must be mapped and the maps integrated to arrive at a reasonable and accurate subsurface picture. These include surfaces such as formations, faults, unconformities, and salt. For example, in faulted areas, it is typically the faults that form the structures (e.g., rollovers, fault bend folds, and fault propagation folds). Therefore, to develop a good understanding of any faulted structure, one must analyze and map the faults. To prepare reasonable and accurate fault maps, an understanding of the types of faults that occur in the tectonic setting being worked, use of all the data, and three-dimensional thinking are required. Further discussion regarding integration of faults and structures, pitfalls of ignoring the integration method, and using isolated fault data are presented in several sections of this book. We cannot overemphasize the importance of mapping faults and integrating them with the structure to arrive at an accurate interpretation. *If you want to drill more than your share of dry holes, ignore the need for fault surface maps, and use some rule of thumb such as the "Rule of 45" (Tearpock and Bischke 1991).*

5. The mapping of multiple horizons is essential to develop reasonably correct, three-dimensional interpretations of complexly faulted areas. The mapping of multiple horizons combined with a series of *Problem-Solving* cross sections allows the mapper to establish a reasonably correct three-dimensional structural framework, prior to generating prospects.

The mapping of multiple horizons (at least three: shallow, intermediate, and deep) provides the mapper with an interpretation that is plausible and fits at all levels from shallow to deep horizons. Remember, almost any set of fault and structural data can be **forced** to fit on one horizon. The true test of the interpretation is to have the data fit at all structural levels. *Therefore, prospects should be presented with at least two or three mapped horizons.* This multiple horizon mapping insures that the interpreted structural framework is geologically sound, and that it conforms to three-dimensional spatial relationships.

All structural interpretations must also balance in volume or area. Therefore, where possible, cross sections should be structurally balanced. By balancing an interpretation, the interpreter is directed toward a geologically and geometrically accurate solution (Tearpock and Bischke 1991).

6. Accurate correlations (well log and seismic) are required for reliable geologic interpretations. As discussed in Doctrine No. 3, all subsurface data should be used in an interpretation. An interpretation that properly integrates all data, such as well log, seismic, and production data is always more accurate than an interpretation that ignores one of these sources (Tearpock and Bischke 1991). Likewise, the correlations must be accurate, because geologic interpretations have their foundation in correct correlations. Consider that all aspects of subsurface mapping are based on correlations. Some of these are the preparation of cross sections, fault, unconformity, salt, structure, and isopach/isochore maps.

Eventually, a mapper's correlations, right or wrong, are incorporated into the final interpretation. **Incorrect correlations can be costly; they can** result in a dry hole, an unsuccessful workover or recompletion, the purchase of an uneconomic property, or the sale of a producing property that has significant, unrecognized potential.

7. The use of correct mapping techniques and methods is essential to generate reasonable and correct subsurface interpretations. The most accurate geologic interpretations are prepared by mappers who have a good understanding of the mapping methods applicable in the area of study. *There is no substitute for correct mapping techniques.* A poor understanding of mapping techniques can result in incorrect procedures, *unjustified short cuts,* and inaccurate interpretations. Although subsurface maps are the primary vehicle used to prospect for undiscovered hydrocarbons, the teaching of subsurface mapping methods is perhaps the subject given the **least** amount of time in college and company training programs.

8. Interpretive contouring is the most acceptable method of contouring subsurface features. Unlike other contouring methods, such as mechanical or equal spaced, interpretive contouring allows the mapper to use knowledge of the structural and depositional style in the tectonic setting being worked, the ability to think in three dimensions, experience, imagination, and geologic license to generate an interpretation that is geologically sound.

9. All work should be documented. Significant volumes of data are collected, evaluated, used, and manipulated during a mapping project. The documentation of these data makes the mapper's work go more smoothly. It helps the person who is evaluating the work to follow the mapper's logic, independently analyze the data and methods used, and quickly arrive at a decision.

10. Sufficient time and detail is required to generate reliable prospects. *Haste makes waste.* ***Do not be too anxious to drill that next dry hole.*** There are not many shortcuts to good mapping. With limited time available to complete a project, alternate solutions may not be analyzed, all the data may not be used, unjustified shortcuts might be taken, or incorrect techniques may be applied. The possible result is a geologic or economic failure. When you consider the cost of a dry hole, an unsuccessful exploration program, or the loss of investor confidence, the time taken to do the job right is time well spent. **Remember Murphy's Law - "if something can go wrong, it will".**

CHAPTER TWO

DRY HOLE ANALYSIS

INTRODUCTION

In this chapter, we analyze the reasons why many dry holes are drilled and make suggestions on how to improve your success ratio. The following are *five* basic reasons why many prospects end up as dry holes:

1. no structural or trapping conditions found as predicted by mapping;
2. reservoir section absent or tight;
3. objective section lacks hydrocarbon accumulations;
4. discovery lacks a commercial accumulation of hydrocarbons; and
5. mechanical problems (drilling or completion).

The first four reasons for dry holes are concerned with geologic or geophysical problems. These problems are caused by lack of data, poor quality of data, incorrect use of data, or an incorrect geologic or geophysical interpretation. The last reason, results from drilling and engineering problems. A recent study by a large independent indicated that the first four reasons listed were responsible for 97.5% of dry holes drilled worldwide over a five year period. Mechanical problems caused only 2.5% of dry holes for the same period. The percentage breakdown is as follows:

1.	No Structural Trap	17.7%
2.	Poor Reservoir Section	30.8%
3.	Section Wet	35.9%
4.	Noncommercial Accumulation	13.1%
5.	Mechanical Problems	2.5%
	Total	100.0%

In another study recently reviewed, reason number 1. accounted for 35% of dry holes.

FOUR BASIC GEOLOGIC REASONS FOR DRY HOLES

Let's look at each of the first four conditions which can result in a dry hole in more detail. We also analyze each problem and review recommendations for improving your success ratio.

1. *No structural or trapping conditions found as predicted by mapping.* These problems are caused by a lack of both geologic and geophysical data, an insufficient amount of time to generate or evaluate a prospect, seismic velocity problems, lack of good quality data, all the data not being used, isolated data used instead of data integration, incorrect correlations, use of improper models, or a geologically unrealistic or even impossible structural interpretation.

The prediction of structure or trapping conditions should be the least problematic of the five reasons for dry holes. If the data are limited, additional data can be acquired. Velocity problems can be analyzed and velocity gradient maps prepared. The mapper may not know how to apply all the data or may have a short deadline. However, these should not be acceptable excuses for not incorporating all the data in an interpretation. Models and methods exist to aid the mapper in interpreting the geology, but these must be used.

Too often, structural interpretations are prepared without much thought given to the three-dimensional validity of the interpretation. At times, essential mapping procedures are not conducted. Instead, incorrect or unjustified short cuts are taken. In some cases, important geologic surfaces are not mapped and integrated into the geologic picture. Incorrect or unreasonable models may be used to generate an interpretation. Finally, completed interpretations are often required *too quickly,* not giving geologists and geophysicists sufficient time to analyze the data properly, generate a reliable interpretation, or examine alternate interpretations.

~~analyze the data properly, generate a reliable interpretation, or examine alternate interpretations~~.

In recent years, we have entered into a new era of prospecting "*the ten minute map*". Due to financial pressures to perform, management often requests drilling locations in an unrealistic amount of time. These "quick and dirty" maps may be acceptable when prospecting for giant elephants; but, in mature areas, they far too often result in poor economic wells or dry holes, even when 3-D seismic survey data are available.

If a technically correct prospecting philosophy is adopted, (such as the one outlined in Chapter 1), better results can be expected. Yes, the initial work may take a little longer; however, the overall project length should be reduced and the prospect success rate increased. One aspect of time reduction is the elimination of redoing and redoing the work as dry holes are drilled.

Much of the "do it quick" philosophy is a holdover from the "Boomtime" days when money was plentiful and technical work was often considered secondary or not considered at all. "Let's test the idea with the drill bit," were the words of the day. Today's economic climate cannot support this philosophy and the resultant dry holes. And, in mature, developed areas where small geologic revelations (minor variations) lead to new discoveries, precision geology is required to maximize hydrocarbon discoveries and minimize dry hole costs (Brown 1982).

In order to help reduce the percentage of dry holes resulting from lack of structure or trapping conditions, an explorationist/exploitationist should:

a. be given sufficient time to complete a project,
b. have access to data required to study the area,
c. use proper models and existing technology,
d. have a good understanding of the structural geology or access to a structural geologist, and
e. generate interpretations that are geologically reasonable and three-dimensionally sound.

Finally, remember there is no substitute for correct mapping techniques. A poor understanding of mapping techniques can also result in inaccurate interpretations.

2.   *Reservoir section absent or tight*. These problems result from prospect objectives being pinched out, shaled out, or absent due to an unconformity or fault. Also, the objectives may be tight.

In the last decade there have been significant advances in sequence ·stratigraphy, and in particular, seismic stratigraphy. These advancements are not the answer to all problems, but they are tools that should be used where applicable. Paleoenvironmental and depositional studies provide a better understanding of local, as well as, regional stratigraphy.

To analyze unconformities and faulting, seismic and well log data must be analyzed in detail. For example, in normal growth fault areas, analysis of fault surface shape and the change in fault dip with depth can often provide important information about the presence or lack of sand at prospective horizons (Bischke and Suppe 1990; and Tearpock and Bischke 1991). The effect of faulting can be further evaluated by preparing fault surface maps within the prospective area.

Diagenetic studies to analyze tight sands are difficult to undertake and time-consuming. However, if core data and well samples are available, work can be undertaken to analyze porosities, cements, and clay content.

The prediction of reservoir-quality rock involves evaluation by geologists, geophysicists, and engineers. Understanding and use of modern stratigraphic methods and depositional models can help reduce the dry holes resulting from the reservoir problems discussed in this section. As in all aspects of a study, proper time must be allocated to conduct the necessary evaluations.

3. *Hydrocarbon accumulation not present.* At times, the geologic conditions found in wells are essentially as mapped, but the objective section is wet. In the United States Gulf of Mexico, for example, the presence of wet traps is a common problem. There are three main reasons for this:

Case 1.   non-sealing faults;
Case 2.   the timing of fault movement and fault growth relative to hydrocarbon migration; and,
Case 3.   lack of source rock or maturation in the basin.

Considering the first case, the sealing nature of a fault often depends on the types of rocks juxtaposed across the fault surface. Sand juxtaposed against sand will often not provide a seal for hydrocarbon entrapment. Faults normally seal when a shale section is juxtaposed against sand, if the fault zone has had secondary mineralization, or if the fault zone is smeared with fault gouge.

We can, for many cases, make the assumption that the juxtapositioning of sands across a fault will not result in a seal to trap hydrocarbons (Smith 1980). Therefore, prospects should be analyzed for juxtaposed stratigraphy. Where prospect closure exists, are permeable beds juxtaposed against impermeable beds providing a seal? Are permeable

~~view of migration and trapping (Allan 1989).~~ With this method three major parameters can be analyzed:

a. style of closure;
b. cross-fault geometry; and
c. juxtaposed stratigraphy across the fault.

With workstations, FAPS software can be used to prepare Allan Sections, analyze juxtaposed stratigraphy, and sealing/leakage potential across faults. Such a study not only defines the trap in terms of seal, but can also provide vital information regarding spill point and potential hydrocarbon column height.

In areas like the United States Gulf of Mexico and Africa's Niger Delta, it is critical that the trapping fault be a growth fault. Also, it has long been recognized that intervals of maximum fault or structural growth are often associated with large hydrocarbon accumulations. Therefore, to evaluate the second case (timing), it is critical to study the growth history of the prospective trapping fault(s) or structure. There are several techniques that can be used to test the structure and evaluate the growth history of the fault(s). These include "Growth Index" (Wadsworth 1953), "Expansion Index" (Thorsen 1963), and "bd/d" (Bischke 1994).

Once again, we come down to the amount of detail required to develop a good structural interpretation and the time available to complete the work. One of the best ways to increase the success rate of discovering hydrocarbons, is to improve one's knowledge of the local geology and the various parameters or techniques available for developing a sound geologic prospect.

4. *Discovery lacks a commercial accumulation of hydrocarbons.* When we talk of noncommercial amounts of hydrocarbons we are referring to those situations in which hydrocarbons are found, but their production is uneconomic. Perhaps the sands are too thin, the formation has a high water saturation, only one objective out of five is productive, the volume of hydrocarbons is too small, or the level of contaminants may be high (H_2S, CO_2, N_2, etc.).

We will always have to P&A uneconomic wells. Some of the problems and their solutions are beyond geologic or geophysical analysis, such as the presence of contaminants or high water saturation. But some are within the realm of geologic, engineering, or geophysical study. Some solutions involve trend analysis, producing field analogies, prospecting in areas with multiple objectives, and the use of all modern methods and techniques available for improving your evaluation of generated prospects.

At times when prospecting, we are successful, not because of good scientific work, but because of *serendipity*. Continued success, however, cannot such as the presence of contaminants or high water saturation. But some are within the realm of geologic, engineering, or geophysical study. Some solutions involve trend analysis, producing field analogies, prospecting in areas with multiple objectives, and the use of all modern methods and techniques available for improving your evaluation of generated prospects.

At times when prospecting, we are successful, not because of good scientific work, but because of *serendipity*. Continued success, however, cannot be achieved through the occasional success by accident. Success requires a commitment on the part of management to provide the **tools** necessary to conduct accurate, scientific investigations and the **time** required to undertake these projects. And we, as earth scientists, must continue to improve our interpretive and mapping skills to provide the highest quality work.

WHAT! ANOTHER DRY HOLE?

Let's not be too quick to drill that next dry hole. Considering the costs involved in doing a project right, it is much less expensive (and professionally less painful) to allow the sufficient time conducting a sound geologic/geophysical/engineering study before drilling, than to explain to upper management or investors that you drilled another $3.5 million dry hole.

We far too often hear these immortal words, "the trap was not present, objectives were shaled out, there was no seal, fault timing was too late, seismic velocity problems were not recognized, objectives were wet, sand was too thin, or the interpretation was incorrect."

CHAPTER THREE

GENERAL STRUCTURE MAP
QUICK LOOK TECHNIQUES (QLTs)

INTRODUCTION

When an exploration or development prospect is prepared, the primary vehicle for illustrating the interpretation and presenting the prospect is the structure map. Because of the dollars involved with each well drilled and the large number of prospects generated, it is important to be able to rapidly scan a structural interpretation for accuracy and three-dimensional validity. Major problems with the structural picture must be identified quickly, so that time and money are not spent on prospects that are geologically inaccurate, unreasonable, or even impossible.

The QLTs presented in this chapter are designed to help you quickly screen the overall structural validity of an interpretation. The QLTs might indicate outright geologic busts or areas that require some additional work to confirm the interpretation. Quick look techniques are powerful tools.

EQUAL-SPACED CONTOURING METHOD

Many prospect maps, whether exploratory or development, are prepared using an equal-spaced contouring method. Of all the contouring methods (Bishop 1960; Tearpock and Bischke 1991), the equal-spaced method often results in the *most optimistic* structural interpretation. This method, which assumes a slope of uniform dip over a general area, results in highs and lows that may be an artifact of the contouring method rather than the actual geology.

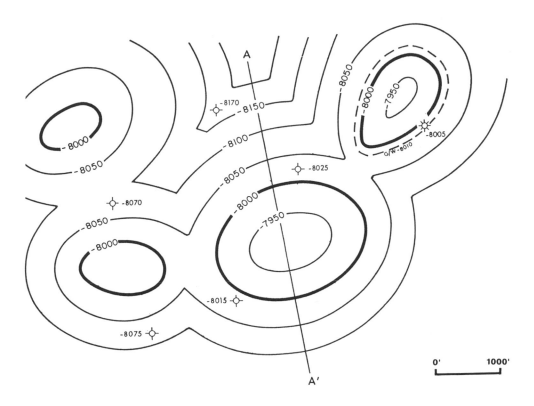

Figure 3-1 Unrealistic structure maps can often occur by using the equal-spaced contouring technique. Observe the sharp changes in contour direction forming contour cusps, and bubble highs. (Published by permission of Prentice-Hall, Inc.)

The use of the equal-spaced method of contouring can result in unrealistic structure maps (Fig. 3-1) with highs looking like *bubble-shaped* structures and lows portrayed as having *sharp cusps*. Compare the two maps shown in Figures 3-2a and 3-2b. Figure 3-2a is a structure map showing three proposed locations updip to existing wells. Notice that the map has been prepared using the equal-spaced contouring method. As shown in Figure 3-1, the equal-spaced method often results in maps showing untested highs adjacent to existing wells (*the elusive high that always escapes penetration by the drillbit*). Thus, the question must be asked, "are the three prospective highs shown in Figure 3-2a real or the result of the contouring method?" Figure 3-2b is an alternate interpretation using the same well data, but prepared using the interpretive contouring method. This method allows the use of **geologic license** to construct the map to reflect the best interpretation for that particular area. Geologic license means that a mapper incorporates experience, understanding of the structural and depositional geology of the area, three-dimensional geometry, imagination, and skill to prepare an interpretation that is geologically reasonable for the area being mapped. Notice in Figure 3-2b that only one prospect is possible, and it may be questionable.

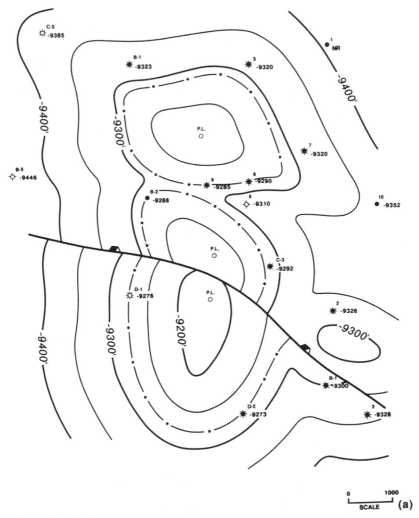

Figure 3-2a The field structure map was prepared using the equal-spaced contouring method. It results in an overly-optimistic structure map. Lay an E-W cross section upthrown to the fault to observe the "Tee Pee" structure.

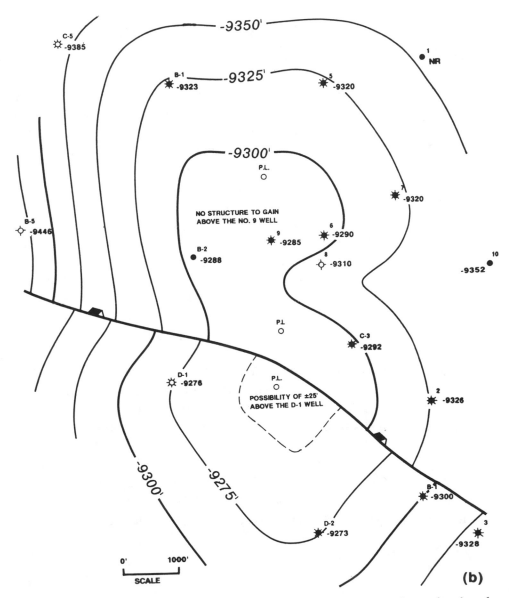

Figure 3-2b Alternate solution for map shown in Figure 3-2a incorporating regional geology and dipmeter data.

Analysis of the method of contouring used to prepare a prospect should be one of the first QLTs applied. If the prospect map is a bubble map, proceed with caution and ask yourself why was the map contoured this way. For a more detailed treatment of the various methods of contouring, we refer you to Tearpock and Bischke (1991).

CONTOUR COMPATIBILITY BETWEEN CLOSELY SPACED HORIZONS

Structure maps prepared on closely spaced horizons should show a configuration that exhibits contour compatibility. Contour compatibility can be easily reviewed by overlaying two maps on a light table. Figure 3-3a shows the overlay of Horizons A and B, which reflects contour compatibility. Figure 3-3b shows the same horizons mapped without compatibility. Notice that in the area of limited control, contours of the same elevation for the two separate horizons cross. This is a physical impossibility.

When closely spaced horizons do not reflect compatibility, the situation may be traced to a number of possible problems including incorrect log correlations, incorrect subsea or deviated hole corrections, over-reliance on questionable seismic data, or plain carelessness. A reliable structural interpretation requires a **three-dimensionally valid structural framework.** The layout of **study aid** cross sections and the mapping of several horizons at various depths helps establish and support a sound structural framework (Tearpock and Bischke 1991).

If structural incompatibility is interpreted in the subsurface from seismic data, it may be correct. There are three primary geologic conditions (Fig. 3-4) that can interrupt a structure with depth and cause structural incompatibility: (1) unconformities, (2) thrust faults, and (3) large listric normal faults.

If any of these events (Fig. 3-5) are present in an area, then the likelihood of incompatibility with depth is great. As a rule of thumb, incompatible contouring, at various depths, should be considered questionable until an independent review of the interpretation is undertaken to confirm the situation. This incompatibility relates to structural, as well as fault, interpretations.

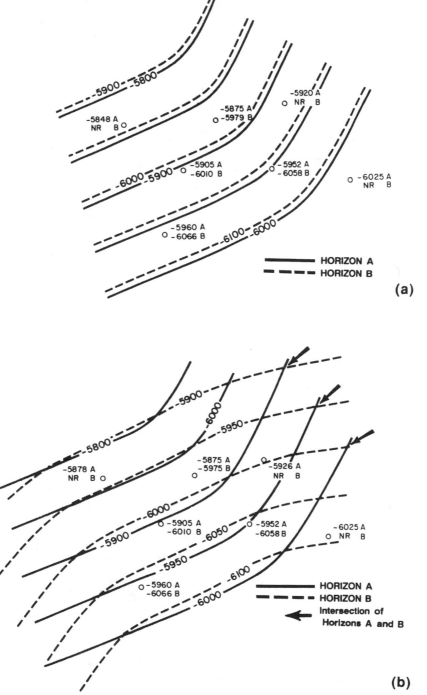

Figure 3-3 (a) Reflects contour compatibility on closely spaced horizons. (b) Reflects non-contour compatibility on same horizons shown in 3-3a. (Published by permission of Prentice-Hall, Inc.)

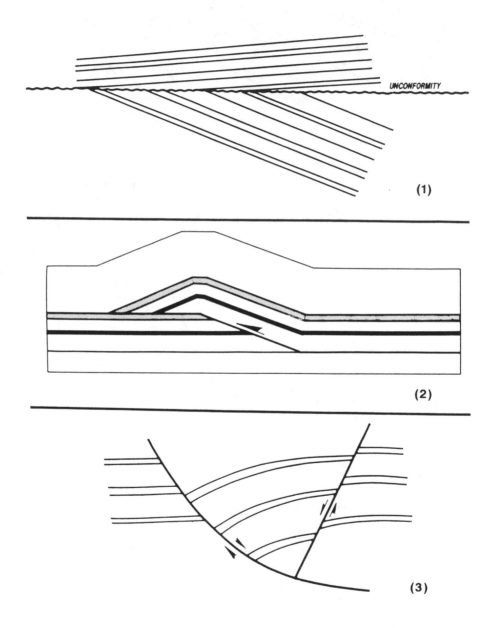

Figure 3-4 There are three main events that disrupt contour compatibility: (1) unconformities, (2) thrust faults, and (3) large growth faults.

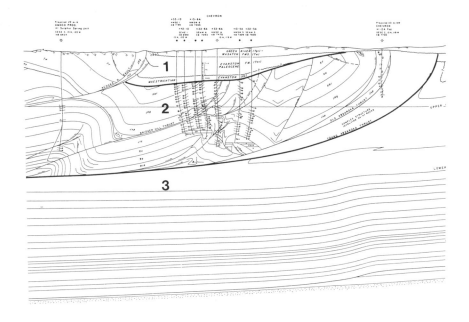

Figure 3-5 Three different structural styles separated by intervening events: (1) Evanston Unconformities and (2) Absaroka Thrust Fault. (Reprinted with permission of Rocky Mtn. Assoc. of Geologists.)

POROSITY TOP VERSUS STRUCTURE TOP

Is the prospect map under review drawn on the actual hydrocarbon bearing sand or formation, or is it mapped on a good marker bed or resistivity marker some distance above the formation? Why are some maps prepared on markers instead of the sand or formation top, and what impact does such a map have on the volume of potential reserves?

Figure 3-6 is a cross section through a reservoir penetrated by three wells. The cross section illustrates a stratigraphic marker which conforms to true structure. The top of a productive sand member is over 100 ft below the marker. Notice that by mapping on the stratigraphic marker above the sand, the hydrocarbon/water contact, which is used as a zero line for the net pay isochore map, is extended far beyond the true hydrocarbon/water contact as mapped on the top of the porous, productive sand member. This added

20

area, created by mapping on the top of the marker, does not contain hydrocarbons; therefore, the volume of recoverable reserves determined from this map will be overestimated. A map on the top of the porous sand resolves this problem. This principle also applies to productive sands with tight intervals near the top of the sand.

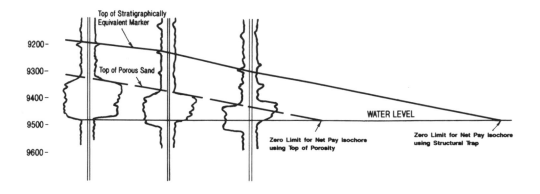

Figure 3-6 A map on the top of structure (a stratigraphic marker) versus a map on the top of porosity will incorrectly define the limits of a reservoir. This incorrect limit or zero line will result in an overestimation of reservoir volume.

Since overestimating reserves by mapping on a marker above a productive formation is a possibility, when should a separate map be prepared on the top of porosity? This decision depends upon at least three factors: (1) the geometry of the reservoir, (2) the thickness of the zone between the stratigraphic marker and the top of porosity, and (3) the relief of the structure. A map on a stratigraphic marker above a sand prepared on a low relief structure will have a *greater* impact on the reservoir volume than one mapped on a steeply dipping structure (Fig. 3-7). The non-productive area included in a reservoir with low relief is larger then one for a high relief structure.

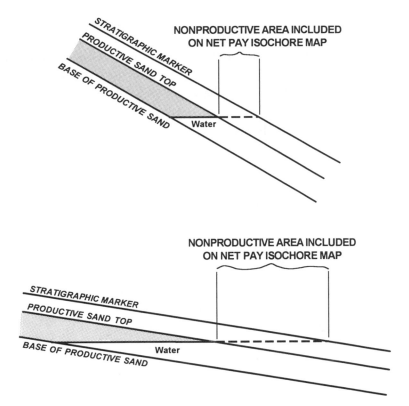

Figure 3-7 The difference in reservoir volume between a map constructed on the top of structure versus the top of porosity is directly related to the dip of the structure. All other things being equal, the lower the dip, the greater the potential error in estimated reserves.

We can not leave this subject without mentioning **porosity bases.** When reviewing prospect maps prepared on multipay zones, one should check on porosity tops as well as checking to see that the estimated reserves have been calculated accurately by honoring the porosity bases. Consider Figures 3-8a and 3-8b, which show two productive Sands, A and B, with a significant shale section between them. Sand A is full of gas to its base, and Sand B has a gas/water contact. There are two ways in which these sands can be mapped. The first method, shown in Figure 3-8a, is to prepare a map on the upper sand and project the gas/water contact from the lower sand onto the map for Sand A.

The second method (Fig. 3-8b) is to map both sands separately and use the lowest known gas from Sand A as the proved gas limit for this sand, and the gas/water contact from Sand B to delineate the proved gas limit for Sand B. The first method provides an unrealistic gas volume because the depth of the gas/water contact in Sand B is not the true limit of gas for Sand A. By using the second method, the base of Sand A becomes the true

22

porosity base for mapping the limit of known gas in this sand. The gas/water contact in Sand B serves as the gas limit for this sand. This method provides a more accurate volumetric estimate of reserves for the two sands. Compare the limit of gas between Figures 3-8a and 3-8b. Porosity bases are also important when a sand is fining downward, where the porosity or permeability of the sand falls below productive limits. In such a case, a map on the base of porosity is more accurate for reserve calculations, than is a map on the true base of sand.

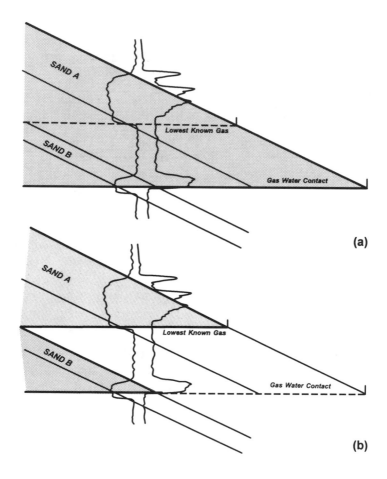

Figure 3-8 (a) Considers both formations to have a common water contact. (b) Uses the lowest known gas for Sand A and the water contact for Sand B.

When reviewing prospects, determine whether the volumetrics are based on a structure map some vertical distance above the porous sand or on a top of porosity map. Don't forget to check the base of porosity as well. This base defines the limit of proved reserves.

RESTORED TOPS AND THEIR USE

One of the **cardinal sins** committed in subsurface interpretation work is the failure to use all the data. The task of coming up with a valid and reasonable structural interpretation is very difficult. The data are limited, and in many cases, the structures being mapped may be miles below the surface. Therefore, to not use all the available data limits one's ability to prepare accurate subsurface maps.

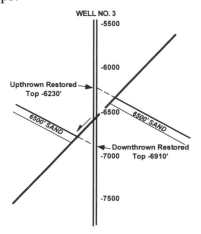

Figure 3-9 The two restored tops for the 6500-ft Sand are -6230 ft for the upthrown restored top and -6910 ft for the downthrown restored top. The difference in restored tops (680 ft) must be equal to the missing section based on log correlations. These restored tops should be used to complete the structure map, for the 6500-ft Sand.

Restored tops are valuable mapping data that should not be ignored, particularly in areas of limited well control. A *Restored Top* (Fig. 3-9) is an estimated top for a specific marker or formation that is faulted out in a well. In other words, *a restored top is an estimate of the depth the top of a formation or marker would have had in a well, if it had not been faulted out* (Tearpock and Bischke 1991). When a formation is faulted out of a well (Fig. 3-9), upthrown and downthrown restored tops provide additional points of control to improve the structural interpretation near the fault. This is particularly true if there is limited well control near the fault (Fig. 3-10) or

24

the strike direction of the structure contours is changing (Fig. 3-11) in the vicinity of the fault.

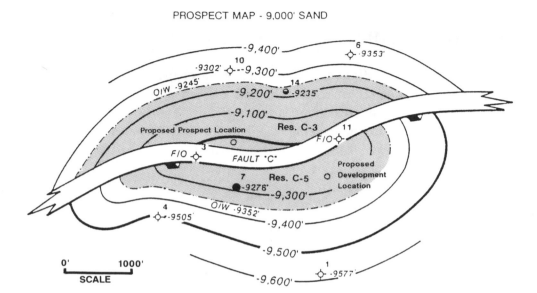

PROSPECT MAP - 9,000' SAND

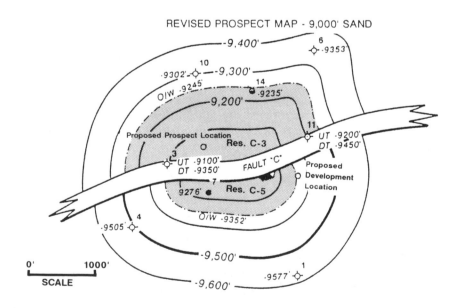

REVISED PROSPECT MAP - 9,000' SAND

Figure 3-10 (a) Completed prospect map constructed without the use of restored tops in Well Nos. 3 and 11. (b) Same area remapped using the two restored tops. (Published by permission of Prentice-Hall, Inc.)

Observe the major differences in the prospect maps shown in Figures 3-10a and 3-10b. In Figure 3-10a, the restored tops for Well Nos. 3 and 11 were ignored; however, in Figure 3-10b, the restored tops were used. Notice the two proposed drilling locations on Figure 3-10a - one in the upthrown and the other in the downthrown fault block. The well control for this interpretation is primarily downdip to the crest of the structure. The only wells that can provide additional data near the fault in the upstructure position have the interval faulted out (Well Nos. 3 and 11). Data from these wells are very important in the preparation of an accurate interpretation. Note how the interpretation changes significantly with the addition of the restored tops for Well Nos. 3 and 11 in Figure 3-10b. The reserves for Reservoir C-3 are reduced by 46% and the reserves for Reservoir C-5 are also reduced, in this case by 42%. In addition to the reduction in volumes, the proposed well for the C-5 reservoir will be a dry hole, based on the more accurate map which used the restored tops.

Figure 3-11, a map of a faulted diapiric structure, illustrates the impact of using restored tops where the structure is changing strike direction in the vicinity of a major fault. Figure 3-11a shows the interpretation using the restored tops for Well No. 4, in which the formation being mapped is faulted out. Figures 3-11b and 3-11c prospectively show optimistic and pessimistic interpretations prepared without the use of the restored tops. Notice how much variation in contouring can be incorporated into the interpretation when the restored tops are ignored.

If a prospect structure map under review has a well or wells for which the formation is faulted out, and no restored tops are used in the interpretation, there are two ways in which the accuracy of the map can be checked. First, you can go back to the well logs and determine the restored tops for the faulted well(s). Place the restored tops on the map and check the map for contour accuracy. If the contoured map is not compatible with the restored tops, recontour the map and evaluate the new interpretation (Fig. 3-10b).

The second method involves the mapping of a shallower horizon, for which the faulted well is in the downthrown block and has an actual well top. Next, map a deeper horizon, one in which the well is in the upthrown block. The interpreted structure for the prospect horizon should, in most cases, honor the structural interpretations generated for the horizons above and below, as well as honor the restored tops.

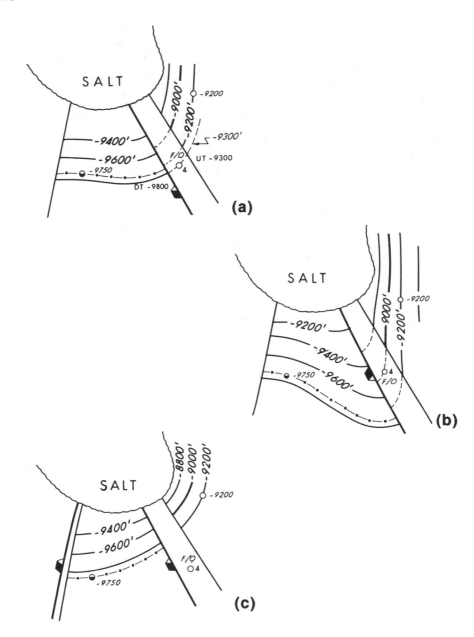

Figure 3-11 (a) Correct structure map using the restored tops in Well No. 4. Ignoring restored tops can result in (b) optimistic, or (c) pessimistic structure mapping.

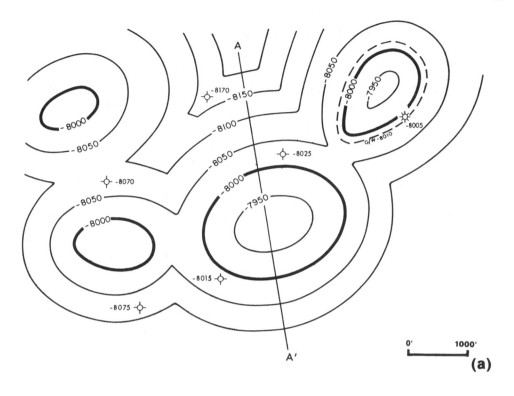

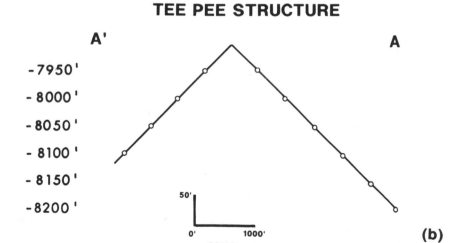

TEE PEE STRUCTURE

Figure 3-12 (a) An unrealistic structure map prepared using an equal-spaced contouring technique. (b) The layout of the illustration aid cross section A-A´ illustrates the unrealistic Tee Pee structure shown in Figure 3-12a. (Figure 3-12a published by permission of Prentice-Hall, Inc.)

THE MAP JUST DOESN'T LOOK RIGHT

There are times when reviewing a prospect map when *something just doesn't look right*. You can not put your finger on anything specific, yet you feel uncomfortable with the interpretation. There is a quick method for checking the overall configuration of a structural interpretation.

The method involves laying out a "quicky" illustration aid cross section (Tearpock and Harris 1987; Tearpock and Bischke 1991) across the structure to review the general structure in cross section. Let's look at the Bubble Map (Fig. 3-12a) which was shown earlier in the text. If we lay a line across the structure as shown on the figure, we can plot in cross section any data that intersects the line. In this case, each structure contour line that intersects the cross section can be used to plot the formation in cross section as shown in Figure 3-12b. The "quicky" cross section shows a **"Tee Pee"** structure, which we know is unreasonable in this case. Be aware that certain structures do have "Tee Pee" shape, including ideal fault propagation folds. Remember, when applying QLTs, the structural style of the tectonic setting must be considered.

Figure 3-13a shows a prospect on two horizons prepared from seismic data. The prospective zone for hydrocarbons is considered to exist between these two horizons. The maps show up to 3000-ft columns of potential gas column. Figure 3-13b shows a cross section of the two horizons with the deeper horizon crossing the shallower horizon. The horizons were reinterpreted and the maps redone. The prospect resulted in a gas discovery in excess of 250 BCF.

If something does not look right, chances are something is wrong. It only takes a few minutes to use the "quicky" illustration aid cross section technique to examine the general overall structural picture for problems. Take time to check any interpretations with which you are uncomfortable. It is time well spent when investment dollars are at stake.

HIGH - LOW'S

The term High-Low refers to a geologic interpretation where a high against a fault has a low contoured against the opposite side of the fault. These are relatively easy to spot on a structure map by examining the relationship of structural highs and lows across a fault.

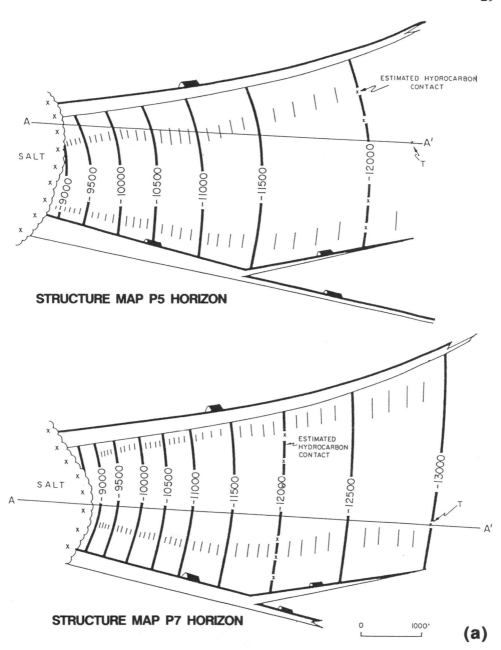

ESTIMATED HYDROCARBON
CONTACT

SALT

-9000
-9500
-10000
-10500
-11000
-11500
-12000

A — A'

T

STRUCTURE MAP P5 HORIZON

ESTIMATED
HYDROCARBON
CONTACT

SALT

-9000
-9500
-10000
-10500
-11000
-11500
-12000
-12500
-13000

A — A'

T

STRUCTURE MAP P7 HORIZON

0   1000'

(a)

Figure 3-13a Structural interpretation of Horizons P5 and P7 based on available seismic data. A-A´ is an illustration aid cross section across the structure. (Daniel J. Tearpock/Richard E. Bischke, APPLIED SUBSURFACE GEOLOGICAL MAPPING, 1991, p. 151. Reprinted by permission of Prentice-Hall, Englewood Cliffs, New Jersey.)

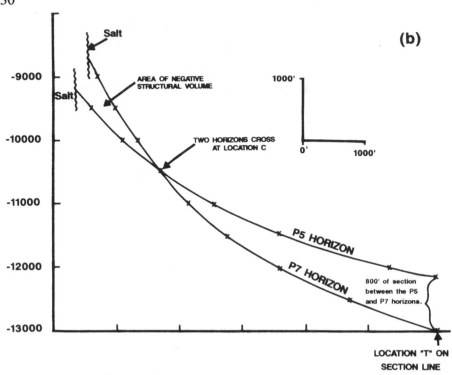

(b)

Figure 3-13b This cross section shows that the lower P7 Horizon crosses the upper P5 Horizon at Location C and is mapped shallower than P5 from this position updip to the salt. This is an impossible interpretation. (Tearpock/Bischke, APPLIED SUBSURFACE GEOLOGICAL MAPPING, 1991, P. 151. Reprinted by permission of Prentice-Hall, Inc.)

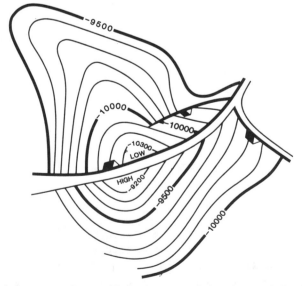

Figure 3-14 High-Low mapping problems in an extensional tectonic setting.

Figure 3-14 is an example of a high-low mapping problem. Observe that a low exists directly across the fault from a high. The structure is in an extensional tectonic setting, and the fault is a listric growth fault. The downthrown low was presented as a rollover structure. This is an unlikely interpretation, since the strike direction of the downthrown contours is at a right angle to the fault. For a rollover structure, the contours would close in the low and strike parallel or subparallel to the fault.

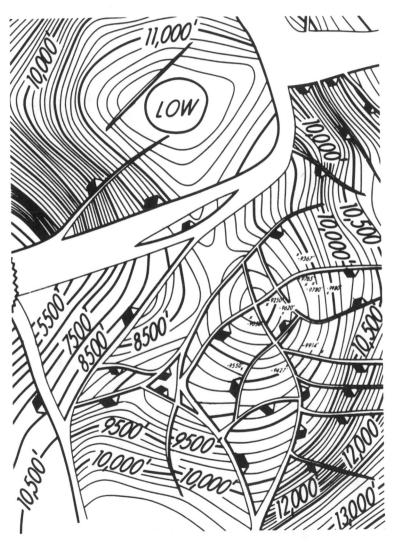

Figure 3-15 This structure map has numerous errors. A contouring error lies in the horst block in the center of the mapped area. Can you find it?

Consider the horst block in the central position of the map shown in Figure 3-15. Following the contours from -8500 ft, try to get to the -10,500-ft contour. Is it possible? The answer is no. The problem stems from running a low across the fault to the west, into a high to the east, a high-low. The fault block is contoured incorrectly and probably interpreted incorrectly, as well.

There are, however, a few exceptional geologic situations where the existence of a high-low is possible. In a salt basin, withdrawal of salt in one block could cause subsidence resulting in a high-low. In compressional basins, a high-low can develop in association with compartmental deformation as a result of a finite compartmental fault (Bell, 1956). And, high-lows are possible with wrench faults.

On a tectonic scale, large normal fault systems can result in high-lows due to isostatic uplift of the footwall by tectonic unloading. The result is that the deepest parts of a half-graben are opposite the highest parts of the footwall. However, under normal geologic conditions, a high directly across a fault from a low should be considered as suspect until the interpretation can be verified.

CHAPTER FOUR

FAULTED STRUCTURE MAP
QUICK LOOK TECHNIQUES (QLTs)

INTRODUCTION

Faulted structures play an important role in the trapping of hydrocarbons. When an interpretation involves a faulted structure, the prospect generator, as well as the people evaluating the deal, must have a good understanding of the geologic and geometric relationships between the faults and structure.

Often, interpretations are made without testing the three-dimensional validity of a faulted geologic structure. Many three-dimensional geologic problems can be quickly recognized by applying a variety of fault quick look techniques. These techniques can rapidly identify geologic and mapping busts and evaluate the effects of the mistakes on the prospect area. Since faulted structures can be complex, we believe that to evaluate a prospect or interpretation and to apply a quick look technique correctly, the evaluator: (1) needs a good understanding of the types of faults that are present in the area of study, (2) should evaluate both the fault and structural interpretation for accuracy and reasonableness, and (3) must be capable of checking whether or not fault/structure map integration has been correctly applied to the final interpretation as shown in Tearpock and Bischke (1991).

FAULT SURFACE ANALYSIS

No prospect or structural interpretation involving faults should be accepted without an interpretation of the important faults in the form of a map. A sound interpretation always begins with an accurate and three-dimensionally valid fault surface interpretation for the tectonic setting being worked. The fault surface interpretation should always be prepared in map view regardless of the well or seismic control and must be prepared with all the available data from logs and seismic sections.

An attendee in a mapping training program once stated, "We don't map faults because they are too complex; we just map the traces on the completed structure maps." If you read this statement carefully, it doesn't make any sense. A fault trace on a completed structure map is the result of the intersection of a fault surface with a formation. Since the final trace is the result of the intersection of two surfaces, it will usually be more complex to define than either of the two separate surfaces. If the faults are too complex to map, then how can one map the traces correctly?

Such a statement may indicate that the mapper: (1) does not understand the importance of mapping faults, (2) does not realize what a fault trace means on a structure map, (3) has not been taught properly, (4) is adhering to an unrealistic prejudice, (5) is repeating the words of someone else, or (6) has the delusion that the three-dimensional subsurface is simple to interpret.

Normal Faults

Let's look at a normal growth fault. Figure 4-1a is an example of a fault surface map presented as part of a prospect package. The prospect is in an extensional growth fault setting. To review the fault for reasonableness, some basic principles of structural geology must be known, and some empirical data for faults must be available.

The setting is an extensional setting like the United States Gulf of Mexico. Many, if not most, of the faults are growth faults - what we term syndepositional faults. Therefore, fault displacement usually increases with depth, and due to sedimentary compaction (see Chapter 7) the growth fault usually flattens with depth, becoming listric (concave up) in shape. We also know from empirical data that the Coulomb shear failure angle - the angle at which rocks fail expressed as a dip angle (or the initial fault dip at the surface) - is approximately 68 degrees.

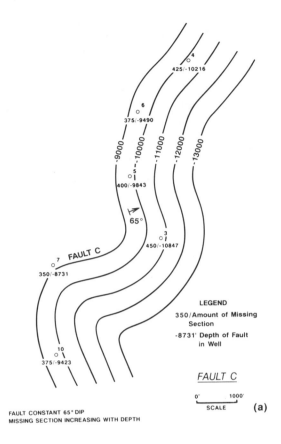

425/-10216

6
375/-9490

-9000
-10000
-11000
-12000
-13000

5
400/-9843

65°

3
450/-10847

7 FAULT C

350/-8731

LEGEND

350/Amount of Missing
        Section

-8731' Depth of Fault
        in Well

10
375/-9423

FAULT C

0'   1000'

SCALE   **(a)**

FAULT CONSTANT 65° DIP
MISSING SECTION INCREASING WITH DEPTH

Figure 4-1a An unreasonable fault surface interpretation of a growth normal fault. Observe the 65 deg constant dip.

With this general knowledge of the area and of structural geology, we can begin to review and evaluate the fault surface interpretation shown in Figure 4-1a. First, notice that there are two right angle bends in the fault surface. At times, such bends may be possible, particularly if the area is affected by cross structures or a transfer zone. Next, observe that the amount of missing section increases with depth. This indicates that the fault is a growth fault. This is confirmed by the prospect generator. Finally, the fault surface is interpreted to have a constant dip of about 65 deg through its entire depth from -9000 ft to -13,000 ft.

From empirical knowledge and structural principles, we know that the incipient dip angle for most United States Gulf of Mexico growth faults is about 68 degrees. Reviewing the fault surface interpretation, we see that the fault is mapped with a constant dip of 65 degrees. Although this dip rate is

36

close to the initial 68 deg, the depth of the fault segment shown is between 9000 and 12000 ft subsea. Since the fault is a syndepositional or growth fault, the dip should have flattened somewhat at these depths from the initial 68 degrees. A review of other well constrained growth faults in this area indicates that, in general, growth faults at these depths dip between 40 and 50 degrees. This constant 65 deg fault interpretation appears to be unreasonable.

Recall from Chapter 3 that we discussed the effects of equal-spaced contouring on a map. When the true dip of a surface is not constant and an equal-spaced method is used, it tends to create sharp angular changes in contour strike direction particularly if the surface is mapped at a higher dip than actually exists. In Figure 4-1a, the two nearly 90 deg changes in strike direction may be an artifact of contouring this fault surface with a constant dip of 65 deg when in reality, the dip is probably lower and most likely not constant at all.

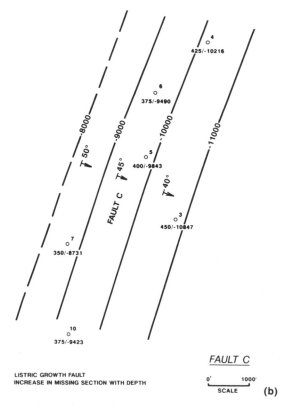

Figure 4-1b A more realistic interpretation of a listric normal growth fault based on the available well control and an understanding of the geological and geometric aspects of growth normal faults.

All these problems lead to the conclusion that the fault surface interpretation is not reasonable for the area and most likely incorrect. Figure 4-1b reflects a revised fault surface interpretation that is more reasonable, compares favorably with other faults in the area, and conforms to compaction principles. This revised interpretation will dramatically affect a prospect mapped in the hanging wall block.

Thrust And Reverse Faults

The analysis of fault shape also must be undertaken when preparing thrust or reverse fault interpretations. Let's look at a thrust fault example. The shape of a fault can affect the shape of the structure, the balancing process, and even the existence and size of a structural trap.

RESTORED

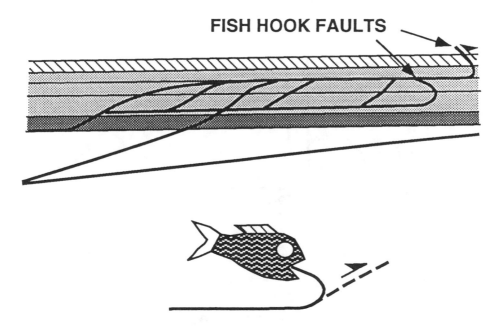

Figure 4-2 Palinspastic reconstruction shows an impossible (fish hook) fault interpretation. See Figure 6-26 for structural interpretation.

Figure 4-2 is a palinspastic restoration of a deformed state cross-section that was supposedly balanced. These restorations are often used to demonstrate that a cross section has been balanced correctly. The cross section comes from an area in the North American Rocky Mountains. There isn't any common, fundamental step-up angle for the thrust faults for each successive ramp (see Chapter 6). The palinspastic reconstruction actually shows an impossible fault interpretation. The fault ramps with the hooks are referred to as *fish hook* faults, which are not possible in this example.

There are other quick look techniques that can be applied to normal or reverse fault interpretations to test their validity. These techniques are presented in Chapters 6 and 7.

STRUCTURAL COMPATIBILITY ACROSS FAULTS

In many tectonic settings the compatibility of structural attitude across faults is more often the rule rather than the exception. Therefore, a faulted structure should have some general structural framework. The use of seismic data is very important in analyzing whether the area being worked or reviewed has structural compatibility across the faults.

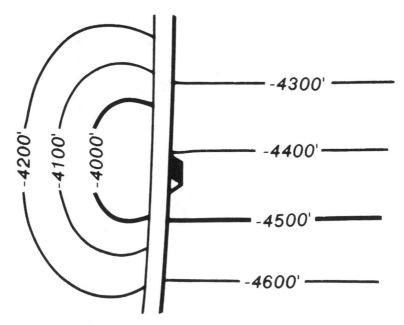

Figure 4-3 Structural incompatibility can often be a red flag indicating an incorrect interpretation.

Structural or contour *incompatibility* across faults can be a result of an invalid three-dimensional structural interpretation, failure to honor all the data, misuse of missing or repeated section, a misunderstanding of the geology, or carelessness. Incompatibility across faults is sometimes easy to recognize, as shown in Figure 4-3, but often the discontinuity is more subtle and difficult to recognize. In such cases, fault cut values are required, as well as an understanding of the correct technique for mapping missing or repeated section across faults (see section on Honoring Fault Data later in this chapter).

The mapping of structural compatibility across faults is particularly significant near the crest of a structure. In this area it may have an impact on well locations and the estimation of potential reserves.

There are five major geologic situations that result in structural incompatibility across a fault (Tearpock and Bischke 1991). These include:

1. large growth normal faults;   **4.** tilted or rotated fault blocks; and
2. large reverse or thrust faults;   **5.** ramps related to rapidly dying faults.
3. wrench faults;

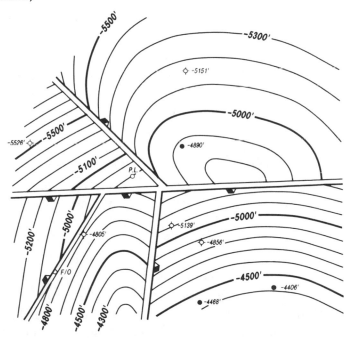

Figure 4-4 Structure map shows contour incompatibility across all faults. Each fault block may have been contoured independently, without regard for the geology in adjacent fault blocks. Five postage stamp maps pieced together. The fault interpretation is also geologically unreasonable.

Structure maps showing incompatibility across faults (Fig. 4-4) are generally incorrect unless the area mapped is affected by one of the five major exceptions. The map in Figure 4-4 shows significant incompatibility across all faults. There are a number of problems with the mapped interpretation, but the *lack of a structural fabric or trend* is by far the most obvious. Other mapping problems apparent on the map, such as improper fault trace construction, restored tops not being used, and an incorrect fault interpretation (e.g., a screw fault), are covered in other sections of the text.

INTEGRATED FAULT/STRUCTURE MAPS

The accurate delineation of a prospect trapped by a fault requires the construction of a fault surface map, a structure map, and the integration (cross contouring) of the two maps (Bishop 1960; Bucher and Hinze, 1962 Tearpock and Harris 1987; Tearpock and Bischke 1991). When working in areas with dipping beds ($\geq 10°$), the position of fault traces and the width of fault gaps (normal fault) or overlaps (reverse fault) on completed structure maps are not intuitive. The fault traces (upthrown and downthrown) on any horizon result from the three-dimensional intersection of two surfaces. It is, therefore, difficult and often impossible *to accurately guess* the position of the resultant fault traces and the width of the fault gap on a completed structure map without integrating the fault and structure maps. This is so, because you are dealing with the geometry of two intersecting surfaces.

In reviewing prospect maps, it is critical to ask if the completed fault trace construction was made with or without the use of a fault surface map. If a fault surface map was constructed, use it to verify the completed structure map. If a fault surface map was not used, a decision must be made as to how much additional work is required to verify the accuracy of the interpretation.

Figure 4-5a is a prospect map illustrating a potential hydrocarbon trap upthrown to a fault and limited downdip by a water contact in Well No.3. Figure 4-5a shows the originally submitted map of a prospective reservoir which was constructed using *isolated fault cut data* from the available wells and two seismic lines to arrive at the completed prospect map. The map shows the formation tops and fault cut data from each well. It appears that the **"Rule of 45"** (Bishop 1960; Tearpock and Bischke 1991) was used to place the fault traces on the completed map. An important question to ask yourself is whether or not the traces are in the correct location. If the upthrown trace is actually further north, a portion of the proposed reservoir area may not exist. Conversely, if the upthrown trace is further south, the reservoir may be larger than mapped.

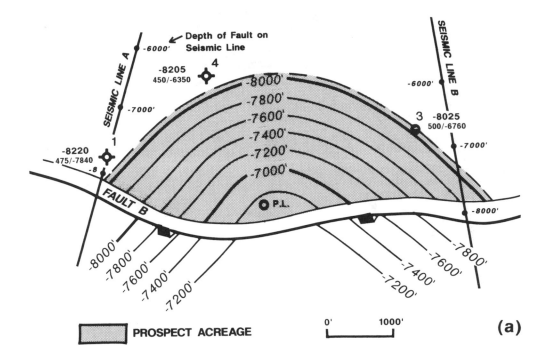

Figure 4-5a Consider the prospect location shown on the structure map. Would you approve this location?

The structure map shown in Figure 4-5b was prepared using the same structural data. In this case, a fault surface map was first constructed (Fig. 4-5c) and then integrated (cross contoured) with the structure map to arrive at the correct positioning of the upthrown and downthrown fault traces. This integration accurately defines the reservoir limit resulting from the fault.

Compare the two maps and consider the new well location proposed on Figure 4-5a. There is a significant change in the position of the fault trace on Figure 4-5b as compared to Figure 4-5a. A well drilled at the proposed location shown in Figure 4-5a would result in a dry hole because the well would be drilled in the downthrown, rather than upthrown, fault block. The enormous error with regard to the positioning of the fault on Figure 4-5a is the result of (1) making the incorrect assumption that the fault dip is 45 degrees, and (2) failing to recognize the effect of bed dip and strike on the positioning of the fault traces over the crest of the structure.

42

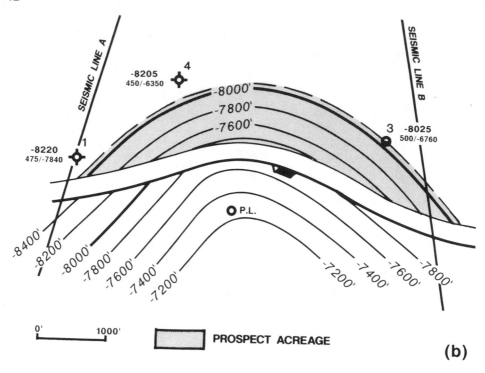

(b)

Figure 4-5b Reinterpretation of Figure 4-5a using the fault surface map for Fault B (Fig. 4-5c) to integrate (cross contour) with the structure map. The fault trace moves approximately 900 ft to the north. A well drilled at the proposed location would actually be drilled downthrown to Fault B.

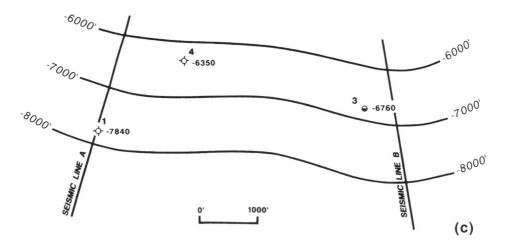

(c)

Figure 4-5c Fault surface map for Fault B constructed from well and seismic control. This figure is not to the same scale as Figures 4-5a and 4-5b.

This error is very common in prospect mapping and has resulted in numerous dry holes. Since we are dealing with a three-dimensional world, the most accurate way of viewing how two surfaces interact in the subsurface is to map each surface and integrate (cross contour) them. This method removes the *guesswork* in attempting to position fault traces.

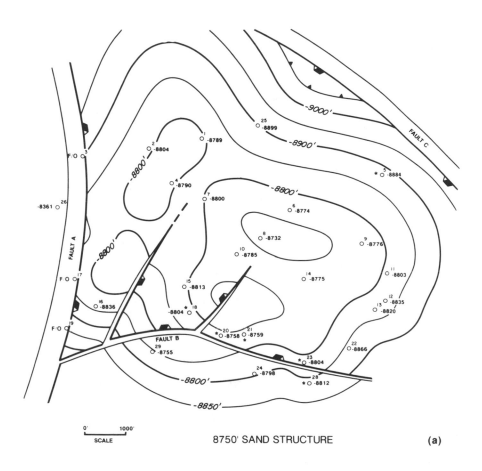

8750' SAND STRUCTURE (a)

Figure 4-6a Structure map on the 8750-ft Sand. Observe the position of Fault B with respect to each well that has an asterisk.

Figures 4-6a, 4-6b and 4-6c illustrate the inaccuracy of fault trace positioning when fault surface maps are not prepared and integrated with the

44

structure map. Consider the position of Fault B on the 8750-Ft Sand Map (Fig. 4-6a) with respect to Well Nos. 18, 20, 21, 23, and 28. Compare the fault position with the same wells for the 9350-Ft Sand Map (Fig. 4-6b) which is ±600 ft deeper. Notice that the fault trace is virtually in the same position on both structure maps, incorrectly portraying a nearly vertical normal fault.

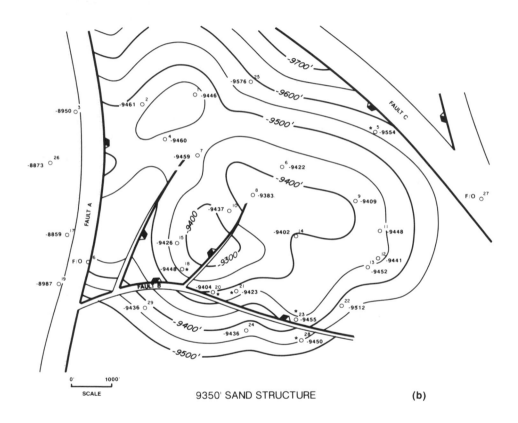

9350' SAND STRUCTURE   (b)

Figure 4-6b The 9350-ft Sand is approximately 650 ft deeper than the 8750-ft Sand. How far north has Fault B migrated from the shallower horizon? What is the dip of Fault B?

Now look at the same fault trace for Fault B on the 9500-Ft Sand Map (Fig. 4-6c) with respect to the same wells. This sand is only about 100 feet deeper than the 9350-Ft Sand, and yet the fault has migrated nearly 700 ft to the north (Well No. 20). The implied dip for Fault B from the 9350-Ft Sand to the 9500-Ft Sand is about 8 degrees. So in this interpretation the fault is implied to be nearly vertical from the 8750-Ft Sand to the 9350-Ft Sand, changing to about 8 deg of dip to the 9500-Ft Sand. As we know, this is an impossible fault picture which tells us that the structural interpretation is wrong. Fault B does not exist as shown on the structure maps.

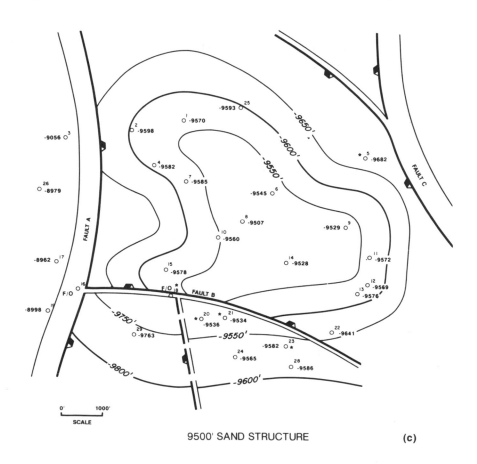

9500' SAND STRUCTURE   (c)

Figure 4-6c Compare the position of Fault B on this map to its position on Figure 4-6b. What is the dip of the fault between the two horizons? Note Fault C, with respect to Well No. 5 on each map.

Finally, look at the position of the downthrown trace of Fault C with respect to Well No. 5 for the same three sands. The trace is 1050 ft from the well on the 8750-Ft Sand, 400 ft from the well on the 9350-Ft Sand, and 800 ft from the well on the 9500-Ft Sand Map. The interpretation indicates that down-to-the-southwest Fault C reverses its dip direction from the 9350-Ft to the 9500-Ft Sand. This is also impossible. Draw this interpretation in cross section, and consider its implication.

Such problems are caused by not preparing fault surface maps to develop a reasonable fault interpretation before proceeding to the structural

46

interpretation. Mapping problems such as these are serious. They must be recognized early and resolved before investment decisions are made.

HONORING FAULT DATA FROM WELL LOGS IN THE CONSTRUCTION OF COMPLETED STRUCTURE MAPS

Normal Faults

The missing section *(Fault Cut)* in the wellbore is what? *Vertical Separation!* I thought it was *throw*.

Does one's understanding of the missing or repeated section, as measured in a well log, affect the construction of a completed structure map? The answer is most definitely yes. It can have a *significant* effect.

Let's first define the term **Missing Section**. It is the true vertical thickness of the stratigraphic section faulted out of a wellbore as a direct result of a normal fault cutting through the section (Tearpock and Bischke 1991). Geologists often refer to missing section as "fault cut". Both are referring to the fault component **Vertical Separation**. For a detailed treatment of this subject, we refer you to the textbook "Applied Subsurface Geological Mapping" by Tearpock and Bischke (1991).

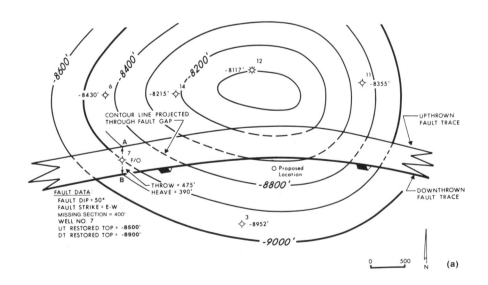

Figure 4-7a Observe how the structure contours are projected through the fault gap from the upthrown to downthrown blocks. (Published by permission of Daniel Tearpock.)

Figures 4-7a and 4-7b show an integrated structure map prepared two different ways. The fault data (Fig. 4-7c) are as follows: fault dip is 50 deg to the south, fault strike is east-west, and the average missing section is 400 ft. The method used to project structure contours through the fault gap for each map is illustrated by dashed contours. Which map is constructed with the correct method for contouring missing section across the fault? The correct answer is Figure 4-7a. It was constructed correctly by mapping missing section as vertical separation. Vertical Separation can be defined in map view as the vertical drop across the fault in the strike direction of the structure contours. Follow the dashed contours in the fault gap in Figure 4-7a. The method used in Figure 4-7b **assumed incorrectly** that missing section is throw (see Tearpock and Bischke 1991).

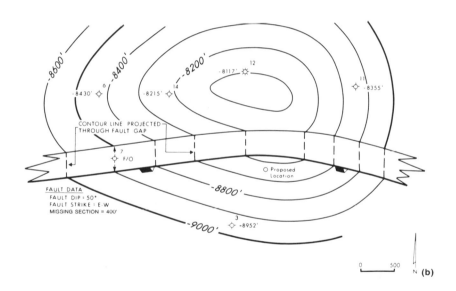

Figure 4-7b This map was constructed using a different method for projecting the structure contours through the fault gap from that shown in Figure 4-7a. This method is incorrect. (Published by permission of Daniel Tearpock.)

Observe on the two maps that the depth at the proposed well location is different. On the incorrect map (Fig. 4-7b) the proposed well location is at a depth of -8640 ft; on the correct map (Figure 4-7a) the depth at the proposed location is -8730 ft. The depth to the horizon is mapped *90 ft shallower* based on the incorrect map. Prospect depth differences of 90 ft can

48

mean the difference between a dry hole and a successful well. This error can reduce the overall structural closure, thereby reducing reserves, or as a minimum, it can result in a well that is not drilled in the optimum position on a structure.

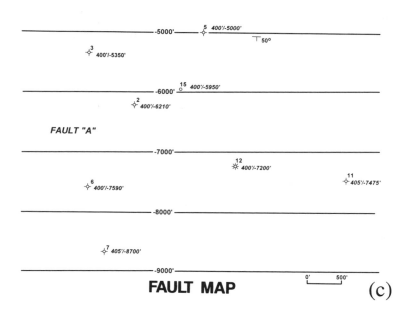

FAULT MAP   (c)

Figure 4-7c Fault surface map constructed from well data.

When vertical separation is incorrectly mapped as throw, **the magnitude of error becomes greater on or near the crest of a structure** (Tearpock and Bischke 1991). This is very critical since hydrocarbons are often trapped on or near the crest of structures.

There are several ways to apply QLTs to verify the correct use of missing section in the construction of completed structure maps. Figure 4-8a is a completed structure map with a prospect in the downthrown block. Notice that there are not any projected contours in the fault gap to indicate how the missing section was contoured. Most mappers do not leave the contours in the gap after mapping across the fault; the contours are typically erased, giving the reviewer no idea as to how the missing section data were used.

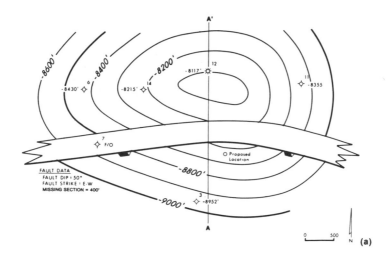

Figure 4-8a Completed structure map showing a prospect downthrown to a fault. The proposed location has an elevation of -8640 ft.

There are three methods for testing the accuracy of the structure map and the viability of the prospect.

1. Project one or more structure contours from the upthrown to the downthrown fault block *along structural contour strike* and calculate the vertical difference in depth as mapped. If the structure map was contoured correctly using the missing section as vertical separation, the vertical difference in structural elevation of a contour from the upthrown to downthrown fault blocks, measured along contour strike, will be equal to the missing section as measured in a wellbore or vertical separation from a seismic line. This is shown by the dashed line in the fault gap (Fig. 4-8b). Graphically, we determine that 300 ft (8700 ft - 8400 ft = 300 ft) of missing section was used in the construction. Compare this value with the amount of missing section identified in the wells, in a seismic section, or indicated by the mapper. In this example, the missing section for Fault A is 400 ft from well log data. **Something is wrong!**

If we project the contours across the fault as if the missing section was mapped as throw (dots in the gap), we get a contoured value of 400 ft (8800 ft - 8400 ft = 400 ft). Therefore, we can conclude that the map was constructed using the missing section incorrectly, since it is the vertical separation (missing section) that is equal to 400 ft and not the throw. The map must be redone.

50

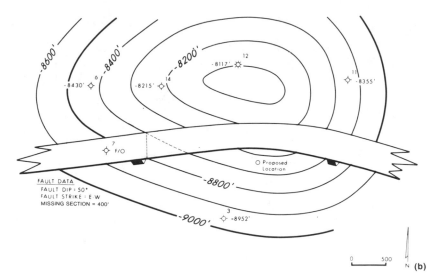

Figure 4-8b The -8400-ft contour is projected through the fault gap from the upthrown block to the downthrown block to determine whether the missing section data were used as throw or as vertical separation.

2. There is a mathematical relationship between vertical separation and throw (Tearpock and Bischke 1990).

$$\frac{(AE)}{(AC)} = \left| \frac{Tan\ \theta}{Tan\ \phi} - 1 \right| \qquad Equation\ 4\text{-}1$$

Where:

AE	=	Vertical Separation (missing or repeated section)
AC	=	True Throw
θ	=	Fault Dip
φ	=	Apparent bed dip measured in the plane of the throw

delete asterisk

θ and ϕ are taken clockwise from 0 deg to 180 deg. The equation can be rearranged to solve for throw.

$$AC = \frac{AE}{\left| \frac{Tan\ \phi}{Tan\ \theta} - 1 \right|}$$

This equation is two-dimensional. There is also a three-dimensional equation which is preferred since it uses true, rather than apparent, bed dips (Sonnat unpublished).

$$AC = \frac{AE}{1 - \dfrac{Tan\ \phi\ Cos\ \alpha}{Tan\ \theta}} \qquad Equation\ 4\text{-}2$$

$\phi =$ True bed dip.

$\alpha =$ Δ azimuth between bed dip direction and fault dip direction.

Equation 4-2 can be used to check the completed structure map in Figure 4-8a. First, graphically calculate the throw on the completed map near the crest along Profile A-A´. Calculate the throw across the fault where the -8200-ft contour in the upthrown block intersects the upthrown fault trace. To do this, a fault surface map must be prepared, since throw is measured *perpendicular to the strike of the fault surface itself* and **not** *to the fault traces* on a completed map. The fault is striking east-west (Fig. 4-7c), therefore the graphical calculation for throw, based on the completed map, is determined in a north-south direction across the fault. The throw as calculated from the map is:

> Throw = 8600 ft (downthrown) - 8200 ft (upthrown)
> Throw = 400 ft

After graphically calculating the throw, use the available data to mathematically calculate the throw at the same position on the map.

Data:

Fault Dip (θ)	=	50 deg
Bed Dip (ϕ)	=	18 deg
Δ Azimuth	=	28 deg
Missing Section	=	400 ft

$$Throw \ (AC) = \frac{400 \ ft}{1 - \dfrac{Tan \ 18° \ Cos \ 28°}{Tan \ 50°}}$$

$$= \frac{400 \ ft}{1 - \dfrac{0.3249(0.8829)}{1.1918}}$$

$$= \frac{400 \ ft}{0.7593}$$

Throw = 527 ft

The graphical calculation for throw from the completed structure map is 400 ft compared to the mathematically calculated value of 527 ft. This discrepancy indicates that the missing section was not honored correctly in the construction of this map. The structure in the downthrown fault block could be in error by as much as 127 ft. Refer to the correctly completed map in Figure 4-7a and graphically calculate the throw of the fault, where the -8200 ft contour intersects the upthrown trace. Graphically, the throw is about 520 ft, which closely agrees with the value calculated mathematically. This indicates that the missing section was honored correctly as vertical separation in the preparation of this correctly completed map.

In reviewing prospect maps under normal conditions, the values for the missing section, fault dip, and bed dip can be obtained. Once obtained, they can be used with Equation 4-2 to calculate the true throw at any location along the fault. If there is a reasonable agreement between the value for throw graphically calculated from the map and that calculated mathematically, it can be concluded that missing section was contoured correctly in the interpretation. As shown in Figure 4-8b, if missing section is incorrectly mapped as throw instead of vertical separation, the value for throw calculated mathematically will not agree favorably with the graphically calculated value. This disagreement may mean that the prospect structure map should be revised using the data correctly *before* making any decision about drilling a proposed location.

3. Construct an illustration aid cross section (Tearpock and Bischke 1991) across the structure along Profile A-A´ in Figure 4-8a, as shown in

53

Figure 4-8c. The vertical difference between the projection of the structure across the fault to the downthrown block should be equal to the missing section as seen in the nearby wellbores.

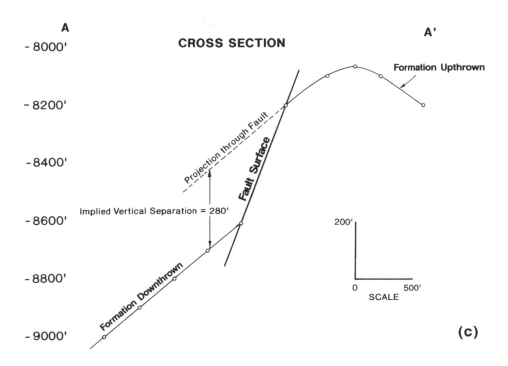

Figure 4-8c Illustration aid cross section constructed from the structure map in Figure 4-8a. Notice that the implied missing section or vertical separation is only 280 ft. It should be 400 ft based on the well log data.

Based on the illustration aid cross section (Fig. 4-8c), the missing section in the wellbores should measure 280 ft. The missing section, as shown on the fault map, is actually 400 ft. Therefore, this cross section technique indicates that the structure map, as completed in Figure 4-8a, is incorrect by about 120 ft.

We have checked the construction of the completed prospect map in Figure 4-8a using three different methods. The potential mapping error for each method was nearly the same.

Method 1 = 100 ft
Method 2 = 127 ft
Method 3 = 120 ft

As an exercise, prepare an illustration aid cross section along the same line in Figure 4-7a, and measure the missing section estimated from this map.

REVERSE FAULT/STRUCTURE INTEGRATION

Reverse faulted structural interpretations are no different from normal faulted interpretations in that a fault surface map should be prepared and integrated with the structure map to generate an accurate interpretation. Balanced cross sections are not a substitute for fault surface maps and structure map integration. Because of the complex nature of how two surfaces intersect (i.e., faults and formations) - especially in highly dipping areas - fault surface maps are almost always required for a reasonable structural interpretation. Finally, the two surfaces should be integrated (cross contoured) to define the position of the fault trace on the prospect horizons.

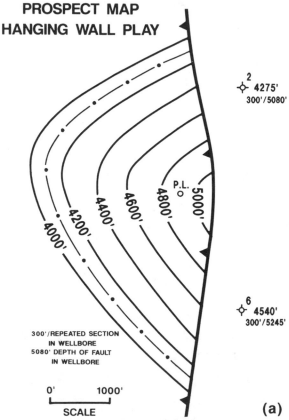

PROSPECT MAP
HANGING WALL PLAY

2
4275'
300'/5080'

6
4540'
300'/5245'

300'/REPEATED SECTION
IN WELLBORE
5080' DEPTH OF FAULT
IN WELLBORE

0'   1000'

SCALE   (a)

Figure 4-9a Hanging wall prospect limited to the east by a reverse fault. Observe the proposed location.

Figure 4-9a is a hanging wall prospect bounded to the east by a reverse fault. Two wells with fault data are shown within the footwall. Evaluate this prospect map noting that the depths are above sea level. After looking at the map for a few minutes review Figure 4-9b which is the fault surface map for the reverse fault. The map is generated from available well log and seismic data. Does the prospect now appear reasonable? The answer is no!

We must always think in three dimensions when dealing with subsurface interpretations. Although the reverse fault is striking north-south, it is intersecting a dipping, curved surface. Therefore, the *trace* of the hanging wall cut off will not be oriented north-south, parallel to the strike of the fault surface. The prospect generator in this case failed to recognize the effect of bed dip and strike on the positioning of the hanging wall fault trace.

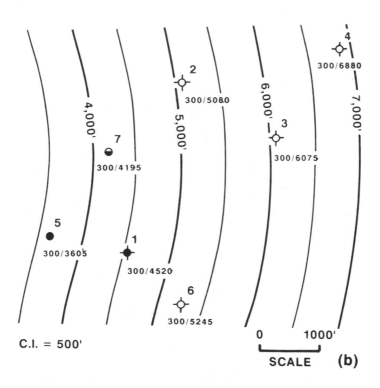

Figure 4-9b Fault surface map based on the well data. (Reprinted by permission of Prentice-Hall, Inc.)

COMPLETED STRUCTURE MAP

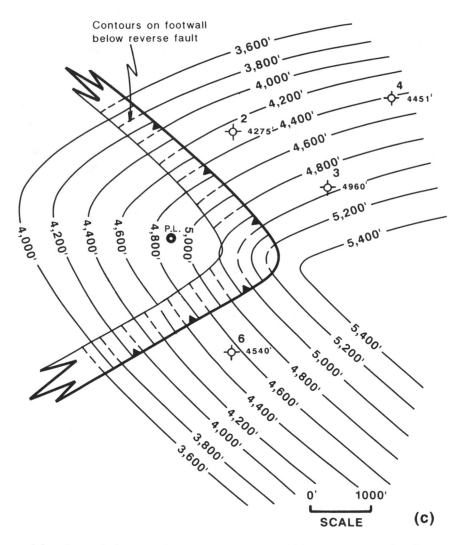

Figure 4-9c Correctly integrated structure map prepared by cross contouring the structure and fault map. (Reprinted by permission of Prentice-Hall, Inc.)

Figure 4-9c illustrates the correct interpretation of the structure and fault trace position. As we showed in the normally faulted prospect, the most accurate way of determining how two surfaces interact in the subsurface is to map each surface and integrate (cross contour) them. The use of this method for mapping prospects removes the guesswork, and accurately positions the fault traces on the final map. The technique takes into consideration the effects of bed dip and strike, as well as fault dip and strike and the amount of missing or repeated section. Remember, repeated section from reverse fault is not fault throw. It is vertical separation. Therefore, when integrating a reverse fault with a structure map, the repeated section must be contoured from hanging wall to footwall as vertical separation and not as throw.

Observe from Figure 4-9c that the proposed location in the hanging wall can be better positioned by moving the location about 1,000 ft to the east to intersect the reservoir in a better structural position.

When a fault surface and a horizon intersect, the defined position of the fault trace for any mapped horizon is not always intuitive nor easy to locate. In Figure 4-9b, the reverse fault is striking north-south. In Figure 4-9a, it was assumed that since that fault strikes north-south, then the hanging wall fault trace must strike north-south. This is not a valid assumption in this or any other case in which the beds have dips in excess of 10 degrees. In Figure 4-9c, we see that the final trace is actually highly curved, giving the appearance of a *boomerang*-shaped fault trace trending northwest-southeast/southwest-northeast. We must always keep in mind that the fault trace on a structure map is the result of the intersection of two separate surfaces, each having a different strike and dip. In this example, it is primarily the changing bed dip and strike across the westward trending nose that causes the curved fault trace.

We again strongly recommend that an exploration or development prospect bounded by one or more faults not be accepted for drilling without the preparation of a fault surface map and the use of the integration technique. Remember, booked reserves are based on volumetric calculations which use the areal extent of the reservoir. Drilling locations are chosen based on fault trace positions.

ADDITIVE PROPERTY OF FAULTS

In an area of intersecting faults, the **vertical separations,** the missing or repeated section, (Tearpock and Bischke 1991) of the individual faults should be additive, or very close to additive, across the intersection where the two faults merge into one. As shown in Figure 4-10, two faults, downthrown to the south, merge laterally to the east with the vertical separation for the resultant fault equal to the sum of the vertical separations for the two individual faults.

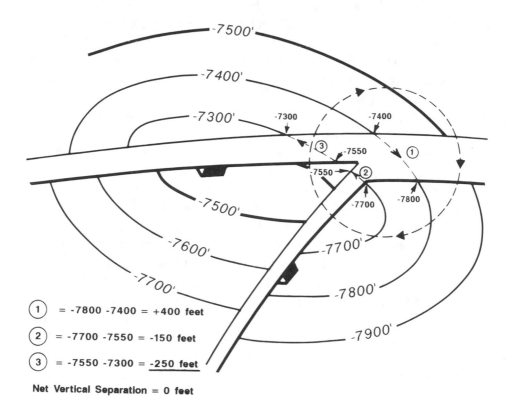

Figure 4-10 The vertical separations for the two individual faults to the west are additive to the single resultant fault to the east.

When checking the additive property of faults, we are not referring to the *fault gap or overlap widths* being additive (see section on Fault Gap Width), but to the vertical separations (the missing or repeated sections seen in the well logs). The vertical separation is checked by calculating the difference in contour values across the fault in the strike direction of the structure contours (Fig. 4-10). An easy way to review the additive property of faults is to go clockwise around a fault intersection (see dashed circle on Fig. 4-10), adding the vertical separations for the faults dipping in a clockwise direction and subtracting the vertical separations for the faults dipping in a counter-clockwise direction. If the additive property of intersecting faults is honored correctly, the additions and subtractions should sum or nearly sum to zero after crossing all the faults. Be sure to check this property as close to the intersection as possible. Faults can change in vertical separation laterally; therefore, at some distance from the intersection the vertical separations may not be additive. Typically, the contours closest to the intersection are used to check this property.

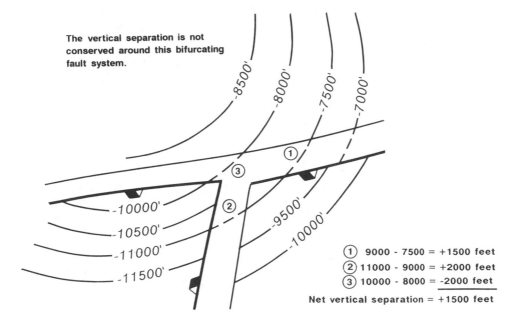

Figure 4-11 The vertical separation is not conserved around this bifurcating fault system.

Figure 4-11 shows a completed structure map prepared from seismic data with intersecting faults that do not honor the additive property. This is a direct indication that the map is incorrect. There is a **1500-ft bust** on this map. Because of the magnitude of this error, it is likely that the problem is the result of a seismic mis-tie across one or more faults.   The data must be reviewed again and the map redone before any decision can be made with regard to hydrocarbon potential.

The additive property of faults is also applicable to small intersecting reverse faults as shown in Figure 4-12. The structure map shows two intersecting reverse faults. Before intersection, Fault A has a vertical separation of 125 ft and Fault B has 150 ft. After intersection, the merged fault has a vertical separation of 275 ft.

We can check the conservation of vertical separation by again measuring the changing contour values across the fault in the strike direction of the structure contours (Fig. 4-12 see points 1, 2, and 3). If the interpretation is correct, the intersection should be in balance and confirm the vertical separation.

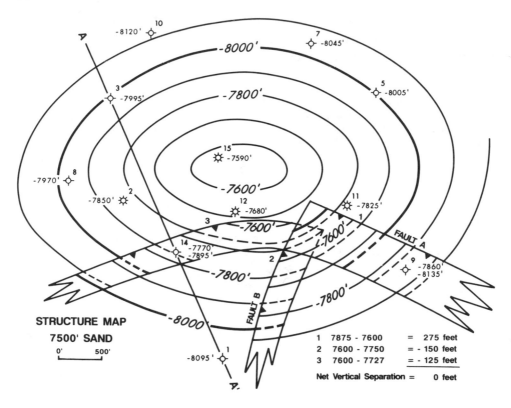

Figure 4-12 The vertical separations across these two reverse faults are additive to the single remaining Fault A. (Published by permission of Prentice-Hall, Inc.)

IMPLIED FAULT ANALYSIS ON STRUCTURE MAPS

Often, a completed prospect map (structure map) is prepared without the benefit or use of a fault surface map. Instead, isolated fault cut data are used to prepare the map. If a prospect you are reviewing has been prepared in this manner, *one of the primary concerns regarding the prospect* would be: are the positions of the fault traces correct? The position of fault traces on a completed structure map are directly related to these factors: the amount of missing section or repeated section, the strike and dip of the fault, and the strike and dip of the formation.

From a completed structure map an **implied fault surface** can be constructed. If the implied fault surface is geologically unreasonable either with regard to dip or fault strike, it follows that the fault interpretation on the completed structure map must also be unreasonable. Therefore, the traces for the fault on the completed map are probably positioned incorrectly.

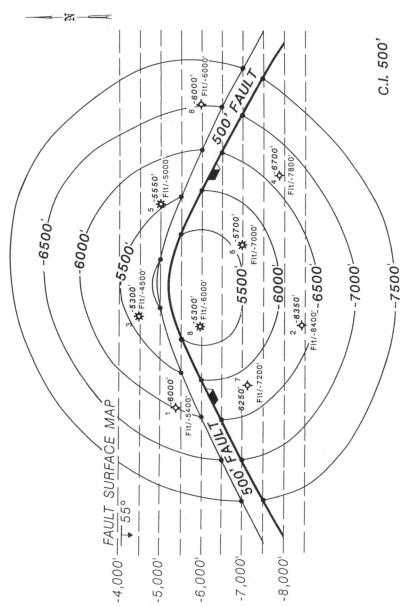

Figure 4-13 When a structure contour of a given elevation intersects a fault trace on a completed structure map, the fault surface must be at the same elevation as the structure contour at the intersection. With a sufficient number of intersections, an implied fault surface map can be constructed. (Prepared by J. Bollick. Published by permission of Tenneco Oil Company.)

This implied fault QLT is powerful. Where each structure contour of a given elevation intersects the fault trace on the completed structure map, the elevation of the fault surface at that point is implied. The elevation of the fault at each intersecting location is equal to the elevation of the intersecting structure contour.

On Figure 4-13, we can see where each structure contour intersects the upthrown or downthrown trace of Fault A. The elevation of Fault A is implied at each intersection point. The compiled fault data can be used to indicate fault strike, and, with enough data points, an implied fault surface map can be prepared. Therefore, if we check the intersection of a structure contour with the fault trace at several locations across the map, the implied fault surface begins to take shape. The black dots in Figure 4-13 denote the intersection of each structure contour with the fault. Each dot shows the implied elevation of the fault along the fault traces. The implied fault surface for this interpretation is shown as dashed lines. Notice that the implied fault surface is quite reasonable, striking east-west and dipping at 55 deg to the south. Paying close attention to the strike of the traces, we see that it is **not** the same as the strike of the actual fault surface. They are not the same because a fault trace is the result of the intersection of the fault surface with the formation.

Figure 4-14a shows a prospect map with potential hydrocarbons trapped downthrown to Fault A. No fault maps were prepared for Faults A or B nor used in the construction of the prospect map. Instead, the position of the fault traces and the width of the fault gaps were estimated based on isolated fault cut data from surrounding wells. The implied fault contours for Faults A and B, determined by the implied fault technique, are shown as dashed lines on Figure 4-14b. The dip of Fault Surface A seems somewhat reasonable at about 42 deg; however, the abrupt change in strike direction may not be as reasonable. Fault B west of the intersection with Fault A looks very unreasonable, dipping at about 83 degrees. Notice the implied position of the -10,000-ft and -11,000-ft fault contours west of the intersection. Obviously, the structural interpretation is not correct, at least with regard to Fault B. The fact that fault surface maps were not constructed during the preparation of this prospect is a problem in itself. With the implied fault surface for Fault B unreasonable for the area, the prospect should be rejected until further geologic work can be done. There are at least three other problems with this interpretation. Can you identify them?

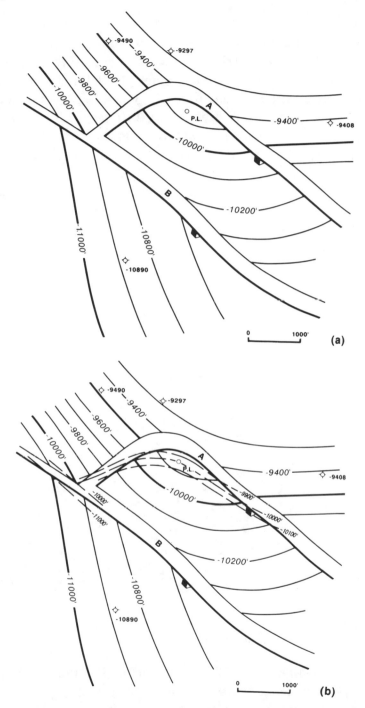

Figure 4-14 (a) Completed structure map with a prospect downthrown to Fault A. (b) Implied fault technique used to construct two (2) implied fault surfaces. What is the dip rate of implied Fault Surface B to the west of the intersection? It is about 40 deg east of the intersection.

64

Look at the unrealistic implied fault from the completed structure map in Figure 4-15. The strike direction of the fault rotates counter-clockwise from -4,100 ft to -4,600 ft through about 150 deg of arc, pivoting around a common center. We call this impossible fault surface a **screw fault**. The fault changes dip direction by 150 degrees over the observed area. At position 1, the fault is downthrown to the northwest. At position 2, the fault is dipping in a northerly direction with no apparent fault displacement. At position 3, the fault is dipping to the east (see further information on screw faults later in this chapter). This is, of course, an impossible geologic interpretation. The fault does not exist as shown on the map and may not exist at all. Perhaps there are two faults.

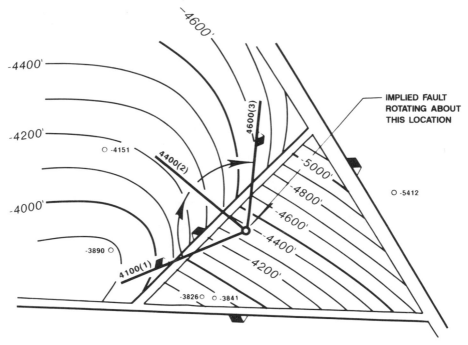

Figure 4-15 The implied fault technique is used to identify an impossible fault interpretation. What is the problem with the fault interpretation?

The implied fault analysis is a powerful QLT that can be employed very quickly to evaluate a fault on a completed structure map. The use of this technique to check the fault picture on a structure map is essential when you know a fault surface map was not prepared and integrated with the final structure map.

THE "RULE OF 45"

The "**Rule of 45**" generally means that in the preparation of a structure map, an assumption is made that the faults in the area being mapped are dipping at 45 degrees. The placement of the final fault traces on a completed structure map will be based on this assumption of a 45 deg dipping fault. Figure 4-16 shows the general application of this rule and how it is used to position a fault on a completed structure map. Figure 4-16 also shows the width of the gap between the upthrown and downthrown traces of the fault. One very important aspect of this rule is often overlooked when applied to mapping. *Not only must the fault be dipping at 45 deg, but the beds must be horizontal or nearly so* for the general rule to reasonably predict the position of the fault traces.

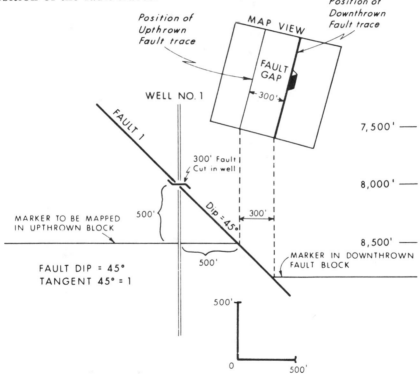

Figure 4-16 The "Rule of 45" can only be applied when a fault is dipping at 45 deg and the beds are flat. (Published by permission of Prentice-Hall, Inc.)

If the fault is not dipping at 45 deg or the beds are not horizontal, the position of the fault traces will be in error on a completed structure map. At times the error can be significant. Consider the example shown in Figure 4-17. The cross section, at the lower left, shows the correct relationship of

66

the well to the fault and the horizon being mapped. The fault cuts the well at -6650 ft, it dips at 60 deg to the east, and it has a missing section of 750 ft. The mapped horizon is dipping to the west at 21 degrees.

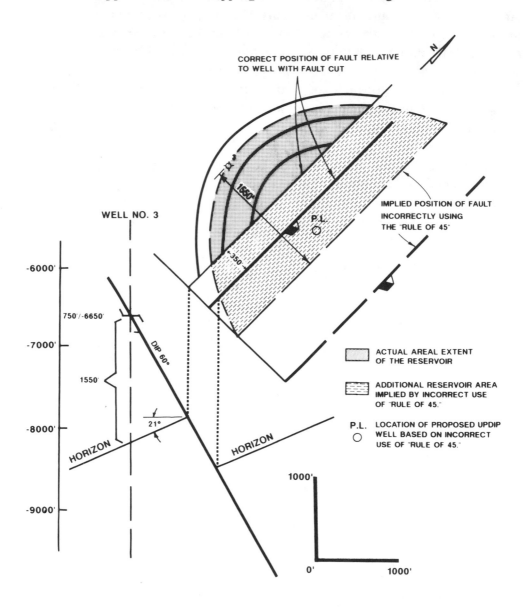

Figure 4-17 The application of the "Rule of 45", when the fault is not 45 deg, or the beds are not flat, can lead to errors in the positioning of the fault traces.

If, instead of preparing a fault surface map and integrating it with the structure map, the mapper working this area decides to use the isolated fault data from the well and the "Rule of 45" to prepare a completed prospect map, the completed map will be in error. The upthrown trace of the fault, implied by using the "Rule of 45", will be placed approximately 850 ft to the east of the actual upthrown fault trace as shown in Figure 4-17. This error incorrectly makes the reservoir appear to be much bigger than it actually is and results in the proposed well location missing the reservoir target. The well will actually be drilled into the downthrown fault block instead of the upthrown block, probably resulting in a dry hole.

The "Rule of 45" is often incorrectly used in prospect generation, particularly in the United States Gulf of Mexico. Where did the idea that faults have a constant dip of 45 deg originate? We don't know, but we can say that, in many cases, it is a bad assumption.

PITFALLS OF FAULT GAP WIDTH

When discussing pitfalls concerning fault gap width, we must discuss several misunderstood ideas about the terms Fault Gap and Fault Heave. Let's define the terms.

1. Fault Gap (or overlap in the case of a reverse fault) is the horizontal distance between the upthrown trace and downthrown trace of a normal fault **measured perpendicular to the strike of the fault traces** (Fig. 4-18).

2. Fault Heave (Fig. 4-18) is the horizontal distance between the upthrown and downthrown traces of a fault **measured perpendicular to the strike of the fault surface itself and not the traces** (Tearpock and Bischke 1991). The strike of the fault surface may not be, and in most cases is not, the same as the strike of the fault traces. The beds must be flat or strike parallel to the fault surface for the strike of the traces to parallel the strike of the actual fault surface.

Notice in Figure 4-18 that there is a significant difference between the strike direction of the "Fault Gap" and that of the "Fault Heave." In this example, the compass direction in which to measure heave is **nearly 90 deg** to the direction required to measure the fault gap. These two fault components are different, and except for specialized conditions, they are not interchangeable.

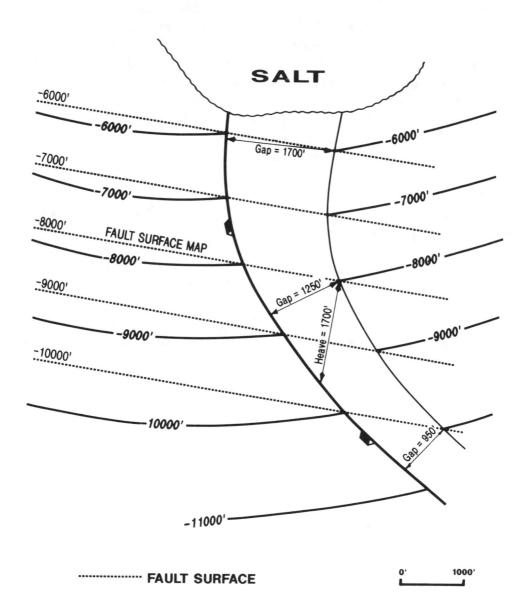

Figure 4-18 Structure map with an overlay of a fault surface map. Notice the difference between fault gap and fault heave. The strike of the fault trace is changing across the mapped area.

There is a widespread misunderstanding that the width of a fault gap or overlap is actually the heave of a fault. From Figure 4-18, we can see that this idea is **not always right** and is, in fact, seldom correct. For the width of the fault gap to be equal to fault heave, the formation being mapped must be horizontal or nearly so, the strike of the fault surface must be parallel to the strike of the formation, and the fault dip must be 45 degrees. In other situations, the width of the fault gap and the fault heave are not the same. Remember, fault gap and heave are measured perpendicular to two different references; fault gap is measured perpendicular to the fault traces, while fault heave is measured perpendicular to the actual strike of the fault surface. The heave of a fault cannot be determined until a structure map is completed or a cross section is made perpendicular to the fault surface. In most petroleum mapping, heave is not as important as fault gap.

There is a *second common misunderstanding* with regard to fault gap. This misunderstanding is based on the belief that the horizontal width of a fault zone measured perpendicular to the upthrown and downthrown fault traces (fault gap or overlap) is equal to the missing or repeated section, as observed from well logs. For example, if the missing section in a wellbore is 500 ft, then the width of the fault gap can be mechanically constructed by scaling off a 500-ft horizontal gap between the upthrown and downthrown traces of the fault using an engineer's scale or 10-point dividers (Tearpock and Bischke 1990). **This is not correct** and can result in maps with significant errors with respect to the width of a fault gap or overlap.

The use of this method for determining the width of a fault gap can only be employed under special conditions. What are these conditions? As discussed in the last section, the fault dip must be 45 deg and the beds must be horizontal. In Figure 4-16, the missing section for the fault is 300 ft and the width of the fault gap is 300 ft. However, in Figure 4-17, the missing section in the wellbore is 750 ft, but the width of the fault gap is approximately 350 ft, which is less than half the value of the missing section. In this example, the fault is dipping at 60 deg and the beds at 21 degrees.

In Figure 4-18, the missing section in the wellbores is 400 ft, but the fault gap ranges in width from approximately 1700 ft measured at the intersection of the -6000-ft contour with the upthrown trace of the fault, to about 950 ft measured at the intersection of the upthrown fault trace with the -11,000-ft contour. *Nowhere in this example is the horizontal width for the fault gap equal to the missing section.* Notice, also in Figure 4-18, that as we move from an updip position, where the beds are dipping at about 45 deg, to a downdip position, where the bed dip is decreasing, the width of the gap is also decreasing and the strike direction of the fault traces is changing. As the bed dip approaches horizontal, the strike direction of the traces will approach the strike direction of the fault. If the fault has a dip of 45 deg and the beds become flat, the width of the fault gap will approach the measured value for the missing section.

Finally, observe that the strike direction of the fault traces in the updip position are nearly *perpendicular* to the strike direction of the actual fault surface. This is not intuitively obvious. On this structure map, you get the illusion that the fault is a radial fault, dying off structure, and dipping to the east. However, the fault is not radial, the missing section is not changing off structure, and the fault is actually dipping almost due south. Faulted structures with high rates of dip or rapidly changing dips can result in very complex geometries.

When checking the width of a fault gap or overlap on a completed structure map, remember that the width is dependent upon a number of factors including:

1. vertical separation (missing or repeated section);
2. fault strike;
3. fault dip;
4. bed strike; and
5. bed dip.

SCREW FAULTS

In the search for screw faults, you don't have to look very far. But, just what is a screw fault? A screw fault is a mapping anomaly having no analog in extensional areas and a questionable analog in compressional areas. It is a fault that laterally reverses its direction of dip. We are not referring to wrench faults that exhibit scissor motion, but to normal or reverse faults. On maps, a screw fault may be portrayed as a normal fault which is contoured correctly (downthrown side lower than the upthrown side) along some portion of the fault, but at some point along the fault it is contoured incorrectly with the downthrown side higher than the upthrown side. In other words, the fault has changed direction of dip.

Consider the M.C. Escher drawing shown in Figure 4-19. Examine the figure carefully. It represents a three-dimensional world drawn in two dimensions. Is the drawing possible in a three-dimensional real world? The answer is no. But, it is illustrated in two dimensions, as if it were possible. In order for the waterfall to exist, the water must run up dip in a never ending stream. This figure is an excellent example of what a screw fault represents on a two dimensional map - an impossible interpretation in three-dimensions.

Figure 4-19 A three-dimensionally impossible situation drawn in two dimensions (© 1961 M.C. Escher/Cordon Art - Baarn - Holland)

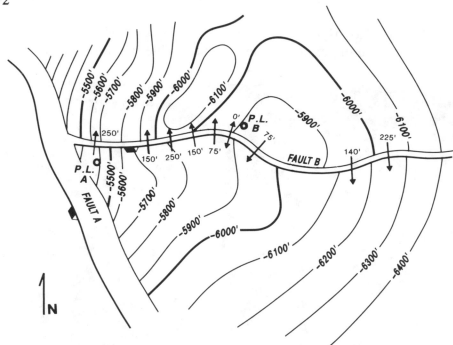

Figure 4-20 Screw Fault B was recognized after the lease was purchased and several dry holes drilled. The fault interpretation is incorrect as mapped.

Figure 4-20 illustrates a screw fault in an extensional setting. At its eastern edge, the fault is dipping to the south with a vertical separation of about 225 ft. Follow along the fault to the west and notice that the downthrown side of the fault is now upthrown by over 250 feet. The fault has changed its direction of dip from south to north. The presence of the screw fault places some real suspicion on the structural interpretation, particularly at the two proposed locations labeled A and B. Is Fault B dipping to the north or south? Is the depth at Proposed Location A -5400 ft? Is it deeper? Shallower? Is there a trapping fault at Proposed Location B?

The recognition of a screw fault indicates some type of error. What is the problem? Perhaps it results from something as simple as careless contouring; however, it may indicate an unreasonable fault interpretation or a correlation bust across a fault. Refer once again to Figure 4-15 where we presented a screw fault. The screw fault indicates a serious problem with the interpretation. Perhaps the fault doesn't even exist. Further analysis is required to evaluate the problem and develop a reasonable reinterpretation.

Figure 4-21a is a completed structure map on the 8300-ft Sand. Look at Fault B at location A. The arrow on the fault trace indicates a normal fault dipping to the south. Follow the fault trace from west to east. Consider area B between the -8300-ft and -8350-ft contours, downthrown to the fault in the eastern region. At this position, the map indicates that the fault changed direction and is now dipping to the north (a screw fault).

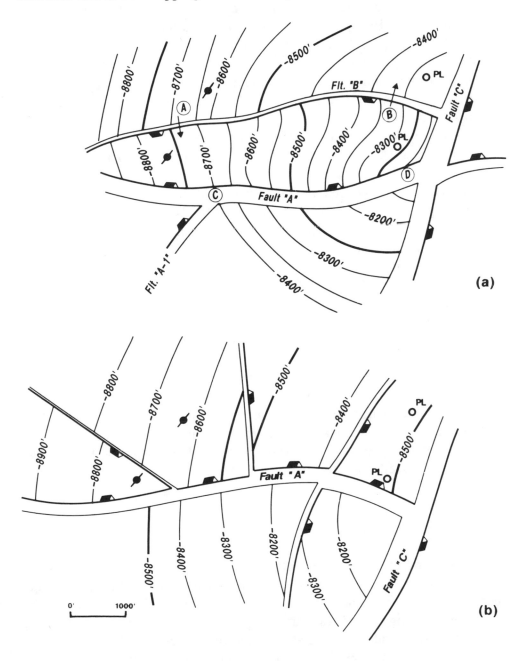

Figure 4-21 (a) Structure map with screw fault. (b) New structure map using fault surface mapping and fault/structure map integration.

Recall, that a screw fault may be indicative of an incorrect fault interpretation. This appears to be what is happening in Figure 4-21a. Figure 4-21b is a reinterpretation of the mapped area, which is significantly different. The reinterpreted map was prepared with the use of data not considered in Figure 4-21a. First, fault surface maps were prepared to develop a reasonable fault interpretation. Second, production data were also used to evaluate volumetric reserves versus performance data in an attempt to further validate the reinterpretation. Third, the faults were integrated with the structural interpretation.

Now look at Location D (Fig. 4-21a) on Fault A. Notice that Fault A has decreased from 300 ft of vertical separation at Location C, to approximately 50 ft at Location D. Fault A is rapidly dying to the east. Also observe that the width of the fault gap for Fault A has not changed despite the vertical separation reducing from 300 to 50 ft. We can say with justified confidence that this map was prepared without the use of any fault surface maps. Remember, it is very difficult to develop a reasonable three-dimensional structural interpretation without first developing a reasonable fault interpretation. The so called *shortcut* of not analyzing the faults has lead to a very unreasonable and more, importantly, incorrect geologic interpretation in this case.

Wells have gone off production in a downdip position (Fig. 4-21) and there are plans to drill updip wells to recover the remaining hydrocarbons based on the map in Fig. 4-21a. The screw fault indicates that the interpretation is wrong. The new interpretation (Fig. 4-21b) shows that the proposed updip locations will not drain all the updip reserves because there is additional faulting that was not recognized from the original mapping.

A map with a recognized screw fault in an area of interest should be rejected until the problem can be resolved.

SAUCER FAULTS

Saucer faults, sometimes referred to as *bathtub* or *stovepipe* faults, are circular or elliptical in shape as shown in Figure 4-22a. This type of fault seems very common on maps and may be a result more of individual mapping style than anything else. However, it can be a result of the incorrect mapping of faults or a misunderstanding about fault kinematics.

A saucer fault results in a mapping bust, as it leads to a problem of mass and space with depth, as well as a misconception of fault geometry. The construction of fault surface maps, their structural integration, and a better understanding of stress relationships should eliminate the incorrect construction of saucer faults. Often a fault, represented as a saucer fault is really two faults forming a compensating fault system (Fig. 4-22b).

SAUCER FAULT

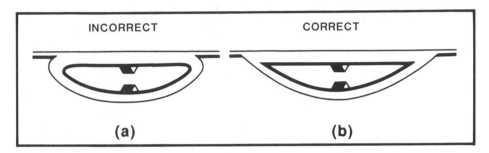

Figure 4-22 (a) Saucer Fault. (b) Compensating fault pattern.

If a mapper honestly believes that a fault such as the one shown in Figure 4-22a exists, there is a misconception about fault geometry. This misunderstanding could possibly lead to a missed opportunity. Consider Figure 4-23a, which is a completed structure map showing an essentially developed and depleted reservoir. Notice that Fault A is a saucer fault turning on itself in the upstructure position. Based on this interpretation the gas reservoir shown is fully developed. The attic potential is insufficient to drill another well. Use the implied fault strike technique for Fault B, in Figure 4-23a, to verify that the fault surface and not the fault trace is implied to make an almost 180 deg turn. This is not possible.

Consider an alternate interpretation shown in Figure 4-23b. In this case, the structure is interpreted to have two faults forming a compensating fault system (Faults A and B). Preparing fault surface maps for both faults and integrating them with the structural horizon results in an updip attic prospect of sufficient size to justify a development well. This example comes from an actual case study where a successful well was drilled for attic reserves.

Except for a few special cases such as collapse of a salt withdrawal basin, karst topography, or syncinally folded thrusts, which can result in a circular fault pattern, these faults should be considered incorrect, rejected as unrealistic, and, if necessary, the area should be reinterpreted.

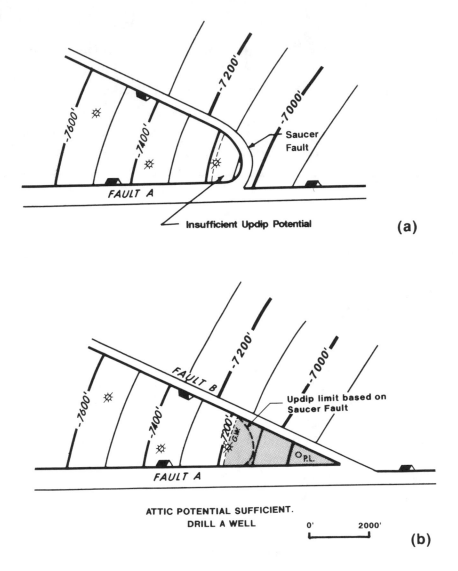

Figure 4-23 (a) Off production gas reservoir bounded updip by a saucer fault. (b) Reinterpretation of fault identifying updip, attic potential.

ODD NUMBER OF CONTOURS TERMINATING AROUND A FAULT

All contours on a continuous surface must close or end at the edge of the map (Bishop 1960; Tearpock and Bischke 1991). This basic contouring rule seems so obvious and simple that no one would violate it. Right? Wrong! Figure 4-24 is a relatively simple structure map with a few faults.

Consider the area east of the major down-to-the-east fault. Is there a contouring problem?

Look at the small down-to-the-west normal fault. Starting at the 10,300-ft contour, try to go around the small finite fault, clockwise or counterclockwise and return to the 10,300-ft contour. Can it be done? The answer is no. Five contours terminate against this finite fault; therefore, one contour is dangling. In other words, one contour does not close.

CONTOUR RULE VIOLATION: ALL CONTOURS MUST CLOSE

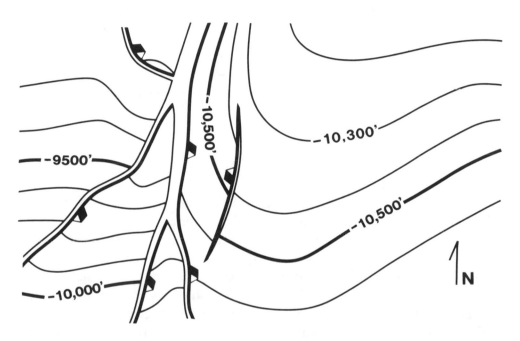

Figure 4-24 The odd number of contours around the small fault indicates a mapping error.

All contours must close (continue across a finite fault); therefore, there must be an even number of contours around a finite fault, not an odd number as shown in Figure 4-24 (Tearpock and Harris 1987). This type of error is very common. A quick way to check a map with a small fault which dies in both directions is to count the number of contours intersecting or terminating against the fault. **If there are an odd number of contours, the construction is wrong.** An even number of contours must intersect a finite fault. This may be a minor mapping bust, but if you find several on one map, it may be time to question the accuracy of the map under review.

CHAPTER FIVE

SEISMIC
QUICK LOOK TECHNIQUES (QLTs)

INTRODUCTION

Even under the time constraints of a meeting, a first pass evaluation of seismic data quality and interpretation accuracy can be made. Seismic QLTs can be applied to evaluate the structural interpretation presented on seismic data. They can be grouped into three main categories based on the types of problems they address. **Data quality and validity techniques** are used to determine if the interpretation has used real data or spurious artifacts generated by the seismic method. **Correlation and mis-tie techniques** look at quality of data utilization by the interpreter and the accuracy of the horizon correlations. **Three-dimensional structural geometry techniques** are used to check the validity of the integration of 2-D seismic line interpretations with 3-D structural map interpretations. (Refer to Chapters 6 and 7 on Structural Geology QLTs.)

DATA QUALITY AND VALIDITY

Data quality and validity problems are generally related to **coherent noise** sources, **incorrect velocities** used in processing, and **bed dip** geometry. Coherent noise events usually appear as steeply dipping events cutting across normal reflectors and have a constant dip wherever they occur on the section. Velocity errors are the primary source of data quality and validity problems. Erroneous structures can be generated by these processing errors. For example, *velocity pull up below a high velocity salt layer can simulate an anticline.* Dipping beds that are oblique to the seismic line can cause out-of-plane reflections or fail to image at all, due to steep dips.

Coherent noise is non-random noise and often appears on seismic lines as lineations of events in sets that are at a steep angle to most of the reflectors (Fig. 5-1; Coffeen 1984). A quick look technique for recognizing **coherent noise** is to use a rolling ruler to compare dipping events. Select an anomalously steep dip that is discordant with the dominant dip on an area of the seismic section. Then roll the ruler across the section coloring events with the same dip. This will prevent you from inadvertently using spurious noise in your interpretations. Once recognized, coherent noise is usually easy to identify on subsequent sections because it often has a single source and thus looks the same on most of the seismic lines in an area.

Velocity errors generate events that are much more difficult to recognize as spurious. The main sources of these errors are near-surface velocity anomalies, anomalous velocity layers, and velocity gradients.

Near-surface velocity anomalies (Fig. 5-2; Tucker and Yorston 1973) are called "statics" problems and result in a vertically aligned time shift of the events directly below the velocity anomaly. *These static correction errors are often interpreted as anticlines that extend vertically over a long depth range, vertical faults, or multiple faults that are vertically stacked.* The quick look key to recognizing static anomalies is the vertical alignment of the shift in the events and the continuation of the shift to shallow depths. An obvious exception to this rule would be a small piercement salt dome, but the surrounding structure would confirm a salt dome interpretation.

Anomalous velocity layers (Fig. 5-3; Tucker and Yorston 1973), such as salt lenses and shallow gas sands, create velocity "pull ups" and "sags" respectively. They can be recognized by the discordant character of the generated structure. Velocity "pull ups" occur vertically below high velocity lenses, and dip increase as a lens thickens. They are common below salt-cored structures and the footwall blocks of faults with significantly different juxtaposed lithologies. *These are common bogus prospects.* The exact thickness of salt must be known to correct sub-salt structures for true depth. Footwall bed dips that increase as they approach the fault surface should be considered suspect. An exception would be complex imbricate fault systems (duplex structures).

80

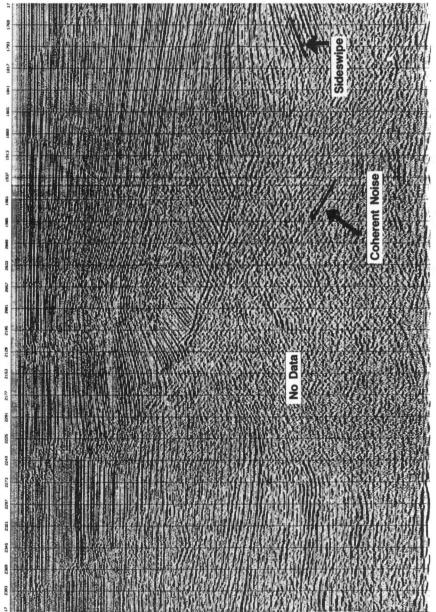

Figure 5-1 Examples of coherent noise event alignments compared to a no data zone (incoherent noise) and side-swipe (out of plane reflectors) on a 2-D seismic line. (Published courtesy of Seitel Data Corporation.)

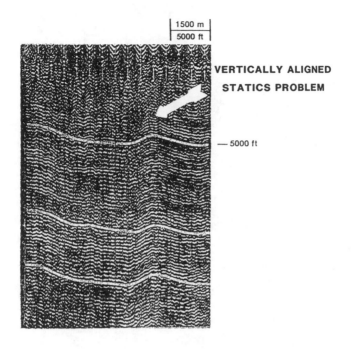

1500 m
5000 ft

**VERTICALLY ALIGNED
STATICS PROBLEM**

— 5000 ft

Figure 5-2 Vertically aligned apparent anticlines due to a shallow velocity anomaly or statics problem. (Published by permission of Society of Exploration Geophysicists.)

Velocity "sags" can be an indirect hydrocarbon indicator, but they distort the deeper structure. They are recognized by the discordant character of the underlying syncline. Typically an isolated syncline is seen below a flat-lying amplitude. Structural restoration of the deeper horizons to accurately map the structure requires detailed seismic modeling of the overlying gas sands.

Velocity gradients can occur laterally as well as with depth. The most common velocity gradient anomalies are beds that appear to thicken onto a structure as shown in Figure 5-4 (Tucker and Yorston 1973). If the change in velocity with depth is great enough, a constant thickness bed will have reflectors farther apart in time on the shallow, low velocity crest of a structure. This will make the structure look flatter than it really is because the syncline will appear too shallow. This is similar to the type of velocity anomaly that obscured the West Chalkley Field in South Louisiana for 20 or 30 years. When finally drilled, the discovery well came in about 800 ft high at the objective. Horizontal gradients cause correlation problems across faults and cause well tie problems. They also cause time migration methods to incorrectly migrate the data. Depth migration is necessary when horizontal velocity differences are present.

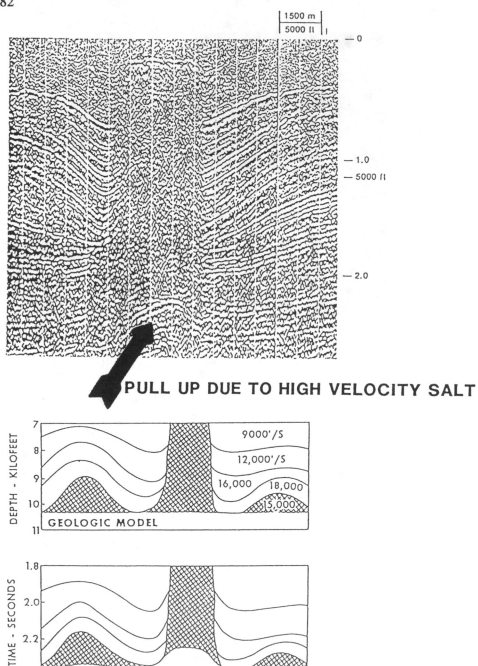

1500 m

5000 ll

PULL UP DUE TO HIGH VELOCITY SALT

Figure 5-3 A high velocity salt plug causing a pull up of the reflectors below it. The large difference in velocity between salt and sediment causes the reflectors below salt to shallow on a time section. (Published by permission of Society of Exploration Geophysicists.)

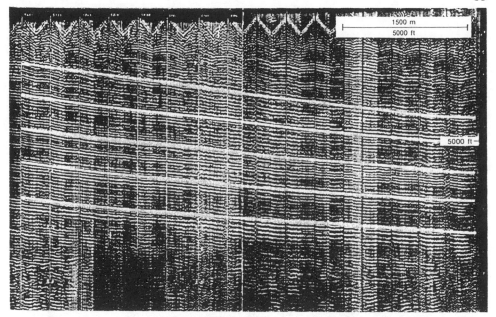

Figure 5-4 Increasing velocity with depth can cause constant thickness intervals to appear to thicken on structure. The same correlation interval has a greater time interval when it is in a shallow low velocity area than when it is in the deeper, and thus, higher, velocity syncline. Faster velocities in the syncline can cause structure mass in time to have anticlines that look flatter than they really are. (Published by permission of Society of Exploration Geophysicists.)

One of the most serious processing errors caused by velocity problems is **incorrect time migration** of dipping beds. This can be difficult to recognize, as it often is associated with **multiples. Undermigration,** as illustrated in Figure 5-5 (Tucker and Yorston 1973), will leave residual diffractions and false anticlines or "bows ties" in synclines. **Overmigration** places reflectors too far up structure and results from using too high a velocity for migration. This can become a major problem in areas of very steep dip ($> 30°$), such as near salt domes, front limbs of rollover structures, or fault propagation folds (Suppe 1985) where most 2-D seismic lines do not image the dipping reflections. Lower dip reflectors can be incorrectly migrated into this area. In the worst case, *multiples from strong, low dip, shallow reflectors* can appear in the "no data" zone at *twice the time (apparent depth) and twice the dip of the primary reflector* (Fig. 5-6). The net result is a false set of moderately dipping reflectors in an area that actually has very steep dip and salt, due to the overmigration of multiples.

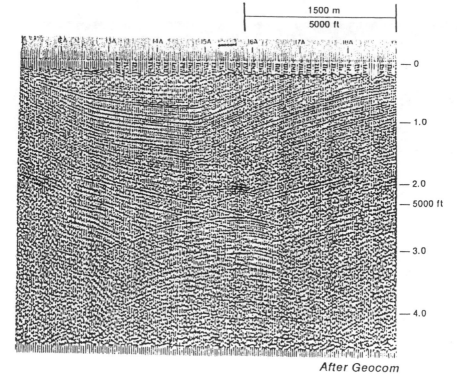

1500 m
5000 ft

— 0

— 1.0

— 2.0

— 5000 ft

— 3.0

— 4.0

After Geocom

Figure 5-5 Using too low of a velocity can result in undermigration with residual diffractions in the synclines that look like anticlines or "bow ties". (Published by permission of Society of Exploration Geophysicists.)

Correct migration can take care of most **bed dip geometry** problems on 2-D lines, except for very steep dips that do not image, and side-swipe. Lines shot close to the strike direction of dipping beds have reflectors that appear too shallow (Fig. 5-7; Tearpock and Bischke 1991). Migrated 2-D dip lines have reflectors that are in their correct positions, thus they do not tie with strike lines. Because reflections return along a line perpendicular to bed dip, as shown in Figure 5-8 (Tearpock and Bischke 1991), they come from a position updip from the plotted position of the line on the shotpoint map. Data points for **strike lines** should be plotted on the map (Fig. 5-9; Tearpock and Bischke 1991) updip from the surface position of the shotpoints. Figure 5-9 also illustrates how to estimate how far updip to post strike line times. Simply slip the strike line along the dip line until the reflector you are mapping ties with the dip line. Post the strike line times this same distance updip from the surface shotpoint positions until the dip or strike of the bed changes. **Side-swipe** (Fig. 5-1) is the extreme case where apparent dips from a structure off to the side of the line are imaged on the line as reflectors cutting across other reflectors or as a discontinuous structure.

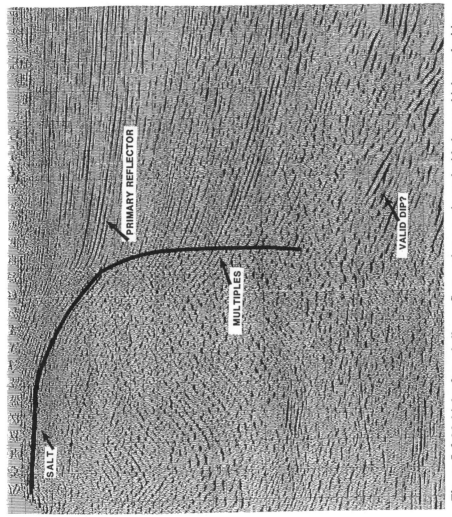

Figure 5-6 Multiples from shallow reflectors that are migrated with deeper, higher velocities are overmigrated and can image in no data zones as if they were valid reflectors. (Courtesy of Coalinga Corporation.)

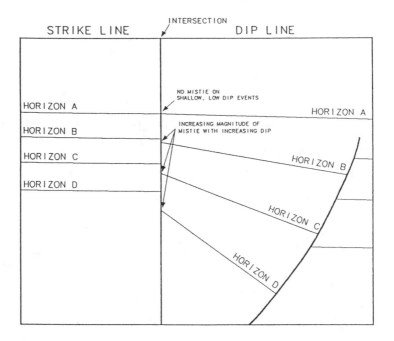

Figure 5-7 Correctly migrated dip lines do not tie dipping reflectors on strike lines at the surface shotpoint location of the intersection of the two lines. (Published by permission of Prentice-Hall, Inc.)

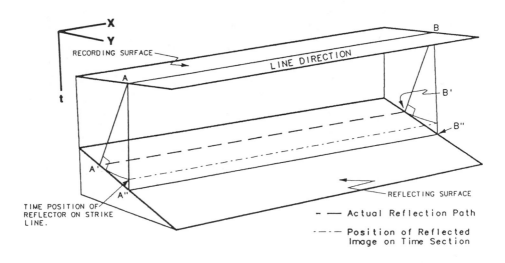

Figure 5-8 The ray path from a reflection point returns to the surface along a line perpendicular to the reflecting surface. For strike lines, reflections are coming from points updip from the surface shotpoint positions on the base map. (Published by permission of Prentice -Hall, Inc.)

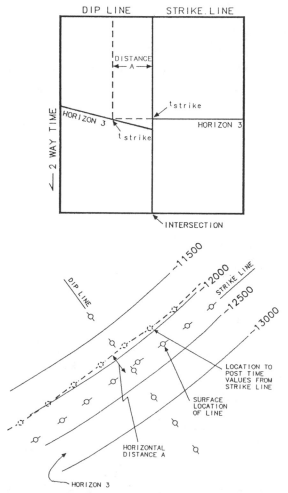

Figure 5-9 To plot strike line times in the correct position, slide the strike line updip until it time ties with the dip line. Move the strike line shotpoints updip this same distance. (Published by permission of Prentice-Hall, Inc.)

CORRELATION AND MIS-TIE TECHNIQUES

Since the objective of seismic interpretation is to make a map that accurately represents the three-dimensional relationships of geologic structures from two-dimensional seismic sections, all the data must be tied together in a grid of closed "loops". **Cross line ties** of mapping horizons and faults are the first things to check when evaluating an interpreter's seismic correlations. Time corrections for lines of different vintages and offset of data for dipping horizons on strike lines should be clearly indicated

88

on the maps and seismic sections. **Wells should be tied to seismic lines parallel to strike of the beds or fault, not perpendicular to the line, unless bed dip is very low or the well is close to the line.** Verify that wells have not been tied to lines in the wrong fault block, such as the example shown in Figures 5-10a and 5-10b (Tearpock and Bischke 1991). This directional well was projected perpendicular to the line, but should have been projected parallel to the strike of the fault surface so it would remain in the same fault block.

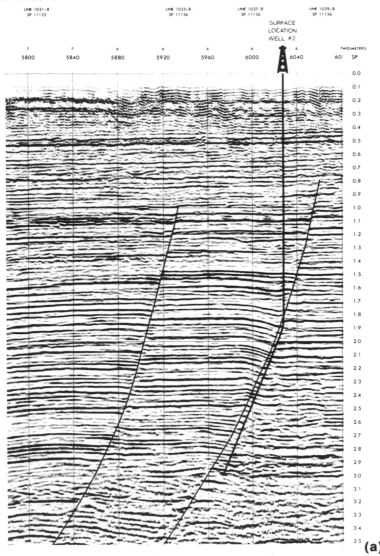

(a)

Figure 5-10a A directional well projected into a seismic line perpendicular to the line appears to be in the footwall block. (Seismic published by permission of TGS/GECO.)

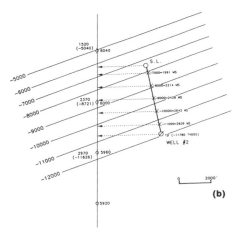

Figure 5-10b A fault surface map shows this directional well is actually in the hanging wall block, but projecting the well perpendicular to the line makes it appear to be in the footwall block. (Published by permission of Prentice-Hall, Inc.)

As a section thickens into a syncline or basin, the increased number of wave troughs can result in the interpreter getting on the wrong "leg" or trough for the horizon pick. *Just as in log correlation, seismic data should be correlated from the thick, complete section to the thin, faulted section.* Check the offstructure well ties first, and then follow the loop ties onto the structure.

Character correlation of groups of reflectors across faults can sometimes be checked by folding the section to line up the correlations in the adjacent fault blocks. Another method is to photocopy a piece of the section, cut along the fault, and slide the two pieces until a correlation can be made. Faults can be correlated from line to line by tying dip lines to strike lines, tying faults that die upward at the same level, or tying faults with the same vertical separation.

On seismic lines, vertical separation can be directly measured. True throw can only be seen on lines that cross the fault perpendicular to the strike of the fault surface, and often the fault must be mapped before this can be determined. As shown in Figure 5-11 (Tearpock and Bischke 1991), you project the dip of the horizon to be mapped across the fault to the adjacent block and calculate the difference in depth using the measured two-way time and appropriate velocity function.

The most common error in identifying faults on seismic sections is to carry the fault pick across *continuous reflectors* (Fig. 5-12a, incorrect; Fig. 5-12b correct; Tearpock and Bischke 1991). This error is often caused by interpreting two faults as if they were one or by forcing a fault interpretation where there is no fault. Also, downward dying growth faults are often interpreted to continue beyond the depth where they die out (see Chapter 7).

90

DETERMINATION OF VERTICAL SEPARATION
FROM SEISMIC DATA

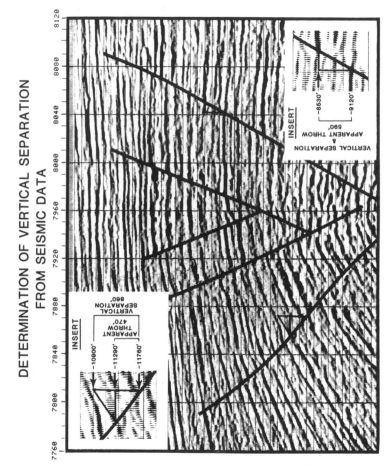

Figure 5-11 Vertical separation can be determined from a seismic line oriented in any direction if an accurate correlation can be made across the fault. The dip of Bed R or Bed G can be projected across their respective faults and the vertical difference in depth of the surfaces can be calculated to determine vertical separation. The difference in depth of the cut-out points on the fault surface is the apparent throw. (Modified from Tearpock and Harris 1987. Published by permission of TGS Offshore Geophysical Company and Tenneco Oil Company.)

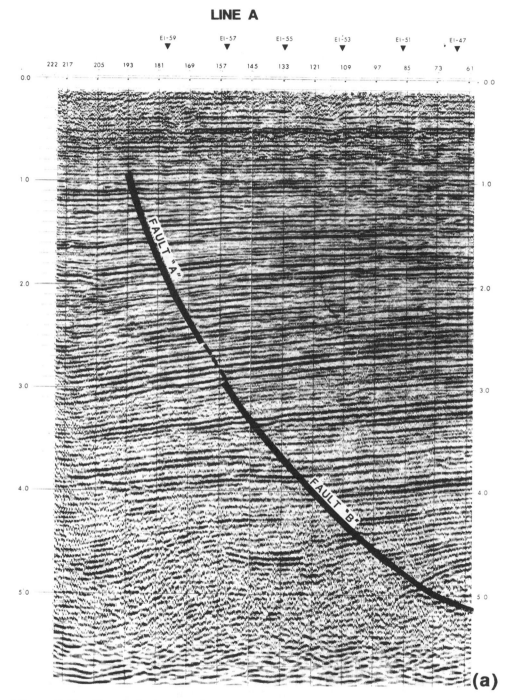

Figure 5-12a Two faults incorrectly connected through continuous reflectors at the dashed line between Faults A and B. (Modified from Tearpock and Harris 1987. Published by permission of Tenneco Oil Company.)

92

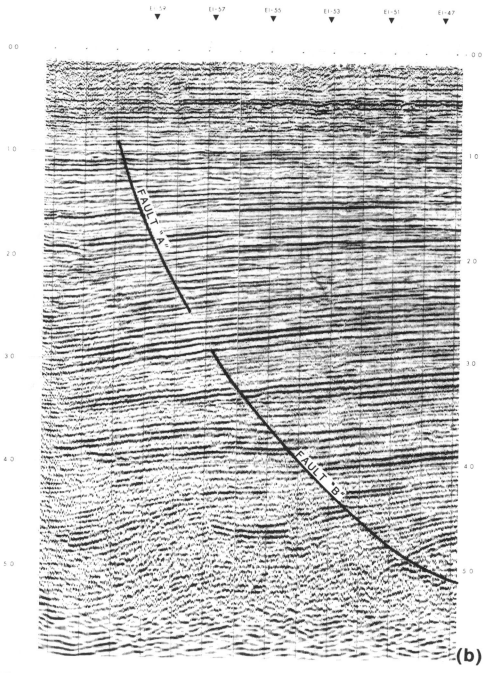

Figure 5-12b Two faults correctly terminated without passing through continuous reflectors. (Modified from Tearpock and Harris 1987. Published by permission of Tenneco Oil Company.)

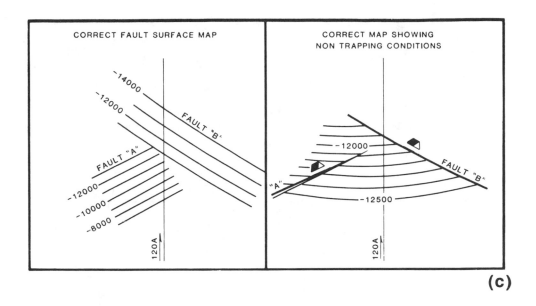

(c)

Figure 5-12c A fault surface map and a structure map constructed, using additional seismic data in the area of Faults A and B, show that the faults are completely different faults. (Modified from Tearpock and Harris 1987. Published by permission of Tenneco Oil Company.)

Figure 5-12a is another example of a two-dimensional interpretation that is impossible in a three-dimensional world. It is another "M.C. Escher" interpretation. Review the Belvedere illustrated in Figure 5-13. It is a simple picture of two porches, one on top of the other. Right? Wrong! If you study the figure carefully you will see that the two porches are at right angles to one another. Observe the positions of the ladder on the upper and lower porches. It is drawn in two dimensions as if it were possible in three dimensions; but, the drawing is not possible in three dimensions.

That is the real problem in trying to portray three-dimensional geometry in two dimensions. The ladder in Figure 5-13 is an analogy of the fault misinterpretation in Figure 5-12b. In two dimensions, the fault interpretation appears reasonable. However, if other seismic lines are used and fault surface maps made, the single fault actually is two faults (Fig. 5-12c). Figure 5-12b shows the seismic interpretation illustrating two separate faults.

Figure 5-13 The Belvedere, a drawing by M.C. Escher of a scene that is impossible in three dimensions. Faults A and B are analogous to the two levels of the structure with the connecting ladder analogous to the dashed line connecting Faults A and B in Figure 12a. (© 1958 M.C. Escher/Cordon Art - Baarn - Holland.)

THREE-DIMENSIONAL STRUCTURAL GEOMETRY

These techniques mainly apply to determining the correct geometry of faults. As a seismic section is two-dimensional and a fault surface is three-dimensional, fault shape cannot be determined, but *can only be inferred from a single section.* Only seismic lines that cross faults perpendicular to strike show true fault dip, and, as many faults change strike laterally, the line may show true dip for the fault at one point and apparent dip at another. This can result in an apparent listric fault on the line (Fig. 5-14; Tearpock and Bischke 1991), even when the dip is constant.

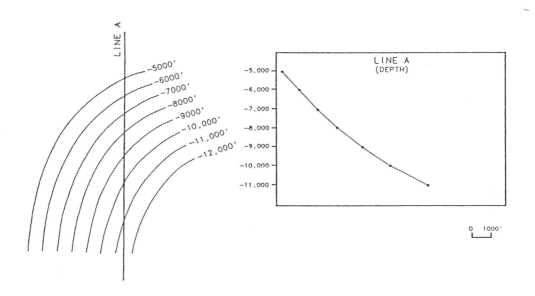

Figure 5-14 A fault with a constant dip rate that changes strike can appear to be a listric fault on a seismic line. (Published by permission of Prentice-Hall, Inc.)

If a fault has been tied across enough lines to determine the strike of the fault, a simple formula can be used to compute true dip from apparent dip as shown in Figure 5-15 and derived from Marshak and Mitra (1988). A nomogram may also be used (Fig. 5-16), which uses the angle between the seismic line and the strike of the fault. But, fault surfaces can image as out-of-plane reflectors on 2-D seismic lines that are not oriented near to the fault dip direction. This method would not work for a fault imaged as an out-of-plane reflector and should only be considered an approximation of the true dip in that case. Often, normal growth faults are interpreted on seismic lines that are oblique to fault strike, with apparent dips that seem plausible for growth faults; however, when the true dip is calculated, it will be much too steep for a growth fault at depth. Also, lines shot at a low angle, between the strike of the line and the strike of the fault, can have problems with side-swipe or out-of-plane reflections.

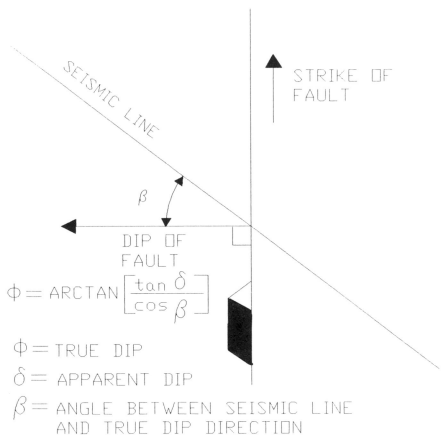

Figure 5-15 A simple equation for recalculating true dip from apparent dip on a seismic line.

There are two methods of projecting the dip of a growth fault for short distances below the last point you can pick on a seismic line. One method uses the amount of drop of a reflector as it rolls into the fault (Chapter 7); the other uses the sand-shale ratio in the footwall block (Xiao and Suppe 1989).

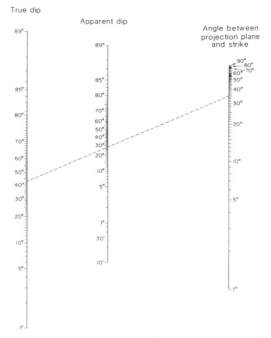

Figure 5-16 A nomogram for calculating true dip from apparent dip. (Published by permission of U.S. Geological Survey.)

Growth fault rollovers deform along Coulomb axial shear surfaces, which can be used to locate the depth to the fault surface and, thus, its dip (Chapter 7). In Figure 5-17, axial surfaces at a Coulomb Shear angle of 68 deg (empirically derived for the United States Gulf of Mexico to be 67.5 deg) are drawn through the inflection or kink points on a reflector; the distance along the axial plane is measured from the depth of the first kink to the depth of the second kink (D_1), and plotted at depth, as shown in Figure 5-17, to get the position of the fault. This only works for true depth sections with no vertical exaggeration. For this graphical method to be applicable a continuous reflector on a time section will need to be plotted in depth with an equal horizontal scale.

Since sand and shale compact by different amounts (Xiao and Suppe 1989), the dip of a buried syndepositional fault will reflect the sand–shale ratio in the footwall block (Fig. 5-18). The quantitative computation of fault dips using the Xiao-Suppe equations is not exactly a quick look technique, but the qualitative application of the method is immediately useful. In zones of increased sand percentage in the footwall block, a fault will tend to steepen. For example, at a 7000-ft depth in the United States Gulf of Mexico, a fault might dip at 45 deg in a 100% shale zone, but in a 60% sand zone, it would dip at 55 degrees.

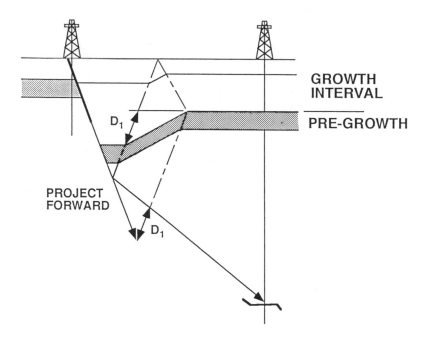

Figure 5-17 A generic rollover model illustrating a depth to detachment method discussed in detail in Chapter 7. Equal distances (D_1) along adjacent axial surfaces can be used to estimate the position of a fault below a rollover.

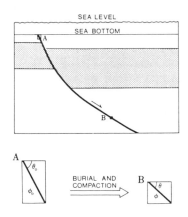

$$\frac{\tan \theta}{\tan \theta_\circ} = \frac{1-\phi_\circ}{1-\phi} \text{ where } \begin{array}{l} \phi_\circ = \text{Initial Porosity} \\ \theta_\circ = \text{Initial Dip} \end{array}$$

Figure 5-18 Growth faults that originate at the surface change dip as the underlying footwall sediments are compacted. (Published by permission of Hongbin Xiao and John Suppe.)

The dip of a rollover into a growth normal fault can also be computed when the fault dips are known (Chapter 7). True fault dips must be used. Using the difference in dip between the upper and lower segments of the fault, the rollover dip angle can be determined using a graph for the Gulf of Mexico Coulomb shear angle (Fig. 5-19). The rollover dip angle can be used to check the mapped rollover closure toward the fault with what the maximum computed dip should be.

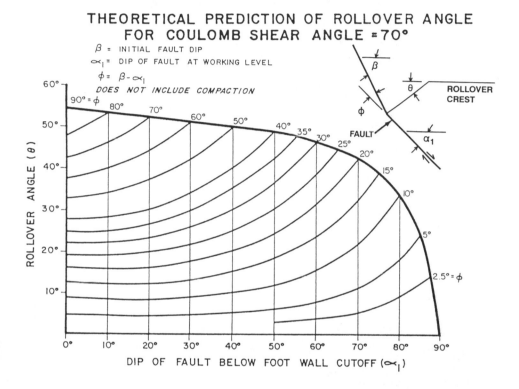

Figure 5-19 A chart for calculating rollover dip caused by a change in fault dip. The dip of the fault below the bend (below footwall cutoff, $\alpha1$) is plotted against the difference in the two fault dips (ϕ). From this intersection point, a horizontal line is drawn to the rollover angle, θ, which is the y-axis on the chart. (Published by permission of Richard Bischke.)

Vertical exaggeration on seismic time lines is a constant problem when attempting to get a quick three-dimensional grasp of the structure. On migrated seismic sections that have not been depth converted, the vertical scale is a function of velocity, and as that changes with depth, so does the scale. One method to help you qualitatively judge fault dips is to plot a 45 deg dip at the depth of interest using a velocity function or survey for the depth interval, and an equal horizontal distance using the shotpoint spacing.

A fault that has an average dip much steeper than this, at depth, is probably not a listric growth fault; or, it may be two separate faults that have been incorrectly connected.

A graph for determining unexaggerated apparent dip from exaggerated dip uses a calculated exaggeration factor (Fig. 5-20; Stone 1991). For a shortened section (vertically exaggerated), divide the length of a horizontally measured unit (e.g., 1000 ft) into the length of a vertically measured unit at the depth of interest to get the exaggeration factor.

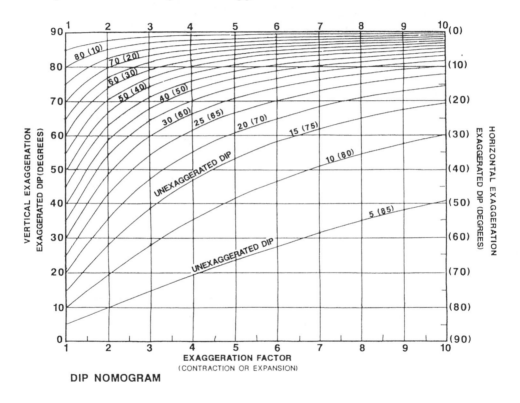

Figure 5-20 A nomogram for calculating unexaggerated apparent dip from an exaggerated dip on a seismic line. The exaggeration factor can be calculated using the group spacing and the velocity at the depth of interest. (Stone, 1991, reprinted by permission.)

QUICK LOOK MAPPING

Geologists and geophysicists commonly use these simple techniques to quickly check the position of large structures in new basins and to determine the plausibility of prospect structural interpretations. After the faults and a horizon of interest have been picked and tied around the loops, the positions of the fault traces, dip vectors, fold axes, amplitudes, and any other important features are plotted on the base map. From these compiled data, faults can be approximately mapped, the structure contoured for gross dip, and the limits of strong seismic amplitudes mapped. Quickly identifying the areas that are most prospective allows you to concentrate your time and seismic budget in the best areas.

All of these techniques can be applied at the conference table or in the office. In just a few hours, you can determine if a lead or submittal deserves the time needed for detailed evaluation and mapping. These techniques are not a substitute for a complete geophysical evaluation by a geophysicist, but should be among the tools a geologist has readily available when shown a prospect based on seismic data.

CHAPTER SIX

COMPRESSIONAL STRUCTURAL GEOLOGY
QUICK LOOK TECHNIQUES (QLTs)

INTRODUCTION

One of the most basic laws of physics that applies to geologic interpretation is that *volume (or matter) must be conserved*; it can neither be created nor destroyed (Goguel 1962). However, if stratigraphic units are miscorrelated or faults misidentified, then a *volume increase (overlap) or decrease (void) is likely to be incorporated into structural interpretations* (Tearpock and Bischke 1991). These interpretations are not balanced (Dahlstrom 1969). Fortunately, *many unbalanced (or incorrect)* interpretations can be identified *prior to drilling* by recognizing possible volume overlaps or voids on existing cross sections. Structural balancing can often prevent the costly drilling of a dry hole, and also reduce the likelihood of having to change the geology every time a well is drilled.

Although volume conservation must be applied in three dimensions, on two-dimensional cross sections, the volume conservation principle can usually be reduced to **conserving area.** In this chapter, we discuss a variety of compressional structural QLTs based on the volume or area conservation principle (Dahlstrom 1969; Suppe 1983; Suppe 1985; Mitra 1986).

A QUICK LOOK AT BED LENGTHS

If temperatures are high, then rock flows in three dimensions and hydrocarbons are volatilized, or cooked out. Rocks within petroleum basins have been subject only to low temperatures and *brittle deformation* (Tearpock and Bischke 1991). Rocks subject to brittle deformation tend to **maintain bed length and thickness within a two-dimensional cross section,** and the concepts of brittle deformation may be employed to maintain cross sectional balance (Suppe 1985). Exceptions to this area conservation (cross section) rule occur across salt domes or shale diapirs. In these cases, material flows into and out of the plane of cross section, and the problem must be balanced in three dimensions.

Figure 6-1a, simplified from Bally (1983), shows a common interpretation of a thrust fold. If one constructs vertical pin lines into the *less deformed portions* of the cover (A,B, and C) and basement (D) units, then these lines provide a reference frame to measure bed lengths. First, select shallow cover Unit B; then select the deepest unit involved in the deformation - in this case Unit C. Take a map wheel or waxed dental floss and measure between the pins lines, the bed lengths on top of Units B and C. Do *not* include the *fault trace* length in these measurements, as the length of the fault trace is not related to bed length. Notice that the bed length on top of Unit B is **much shorter** than the bed length on top of Unit C. Thus, Figure 6-1a does not conserve bed lengths and is geometrically impossible on physical grounds. If there are major busts in the cross section, this procedure can be done in the "mind's eye" with little practice. More subtle problems require a careful analysis.

Figure 6-1b is a reinterpretation of the original non-depth corrected time seismic profile and is in *"rough"* balance. You can check for line length balance by repeating the above bed length measurement experiment. In this reinterpretation, the fault pattern has changed. In Figure 6-1a, a good prospect may have been overlooked by introducing faults into the interpretation that do not exist. In other words, if nonexistent faults are introduced, you may reject the prospect on the grounds that the hypothetical faults provided a pathway for hydrocabons to migrate out of the core of the anticline.

A QUICK LOOK AT BED THICKNESS

Refer again to Figure 6-1a and notice that the thickness of Unit B varies across the fold, thinning across the crest or core of the fold, and then thickening rapidly on the limbs (Figure 6-1a, see arrows). The original seismic section, published in Bally (1983), shows that Unit B is not a growth section and does not onlap Unit C. This unnatural thinning of Unit B suggests that the section is unbalanced.

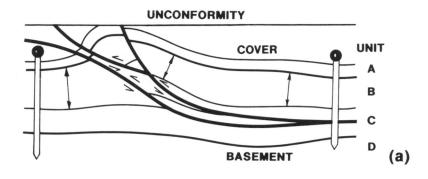

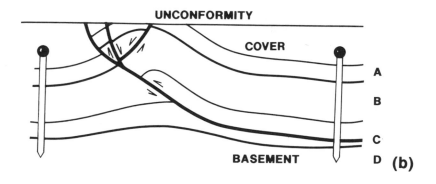

Figure 6-1 (a) Interpretation of an asymmetric fold modified from the literature. Cross section can be shown to be unbalanced by measuring bed lengths along units A to C or by noticing the thinning of Unit B across the top of the anticline (see arrows). The original seismic section, published in Bally (1983), shows that the thinning is not the result of onlap onto the structure, and thus Unit B is not a growth section. (b) Reinterpretation of original seismic non-depth corrected section results in a roughly balanced section. The reinterpretation eliminates several thrust faults, maintaining prospect integrity within the core of the anticline. (Figures 6-1a and 6-1b published by permission of the American Association of Petroleum Geologists.)

By simply observing changes in unit thickness, we can quickly conclude that area is not conserved. An examination of Figure 6-1b shows that this *"roughly balanced section"* maintains unit thickness. Thus, in compressional areas, bed or unit thickness conservation represents one of the simplest and most powerful QLTs.

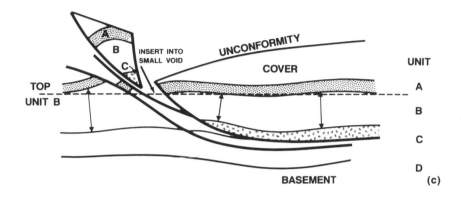

Figure 6-1c Retrodeformed cross section of horizons shown in Figure 6-1a. The retrodeformed horizons can not be aligned to a common level without introducing a volume overlay.

Consider the cross sectional interpretation shown in Figure 6-2a. The cross section shows a listric reverse fault with deformed hanging wall and footwall prospects. Using the bed length and bed thickness QLTs, evaluate the reasonableness of this interpretation and proposed prospects.

Does the cross section balance? The answer is "no" for several reasons. First, by line length balancing, we see that the line lengths increase with depth as shown in Figure 6-2b. This increase in line length with depth has been observed with basement involved structures, and indicates that the shallower cover rocks are thrust off the cross section along bedding plane thrusts.

Second, the bed thicknesses in the footwall change dramatically from the footwall limb to the hinge zone. Although this prospect is in a "thin-skinned" tectonic area of low temperature and low pressure, the footwall structure reflects what resembles high temperature folding, as observed in metamorphic rocks. We would not expect the hanging wall rock to be

106

sedimentary and the same strata in the footwall to be metamorphic when only separated by a small reverse fault. We also know that the strata in the footwall are not incompetent rock flowing into the hinge zone because the same rock units in the hanging wall do not reflect incompetent flow into the hinge. Therefore, strata thickness change into the hinge zone is an unlikely interpretation.

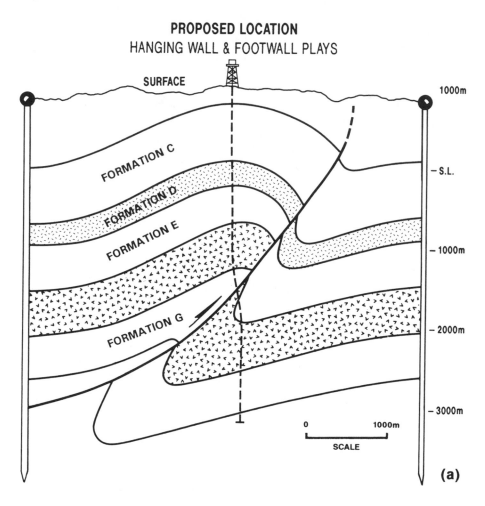

Figure 6-2a Cross section through an oil field located in the hanging wall. Interpretation of footwall is based upon seismic data. Notice the unusual thickening of the footwall beds as compared to the hanging wall beds in the oil field. The proposed well is designed to test the overturned beds in the footwall.

Third, notice how the footwall beds have been significantly folded along the fault surface, creating a footwall closure against the fault. This is a common pitfall reflected in many footwall interpretations. In a number of cases, the turned up beds are not real, but are the result of an incorrect interpretation of velocity pullup. Footwall beds are typically flat unless affected by breakthrough or duplexing (Suppe 1985; Mitra 1986; Tearpock and Bischke 1991). Notable examples of upturned or inverted beds occur beneath the front of some reverse faulted basement structures (Spang et al. 1985; Narr and Suppe 1993).

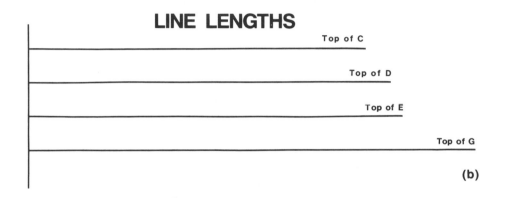

Figure 6-2b Line or bed length measurements increase with depth on Formation C, D, E and G.

Finally, consider the continuously upward listric shape of the reverse fault. The interpretation of a listric reverse fault is quite common in the literature. However, we believe, based on sound structural principles, that most of these continuously upward listric interpretations are incorrect. Areas of high bed dip usually result in no data regions on seismic sections. A listric fault interpretation may mean that no attempt was made to resolve the shape of the thrust fault through no data areas (see section on Thrust Fault Position later in this chapter).

Figure 6-2c is a reinterpretation of the cross section shown in Figure 6-2a using classical balancing techniques. With this interpretation, the bed lengths are the same, the thicknesses are preserved, and the fault shape has been slightly changed. The reinterpretation has resulted in a semi-balanced section, in that the thickness of the shallowest layer is not everywhere constant. The interpretation can be further improved by using the Kink method of cross section construction and the fault propagation fold model

108

(Fig. 6-2d). In other words, the section is balanced using geometric formulas (Suppe and Medwedeff 1990). This procedure is quicker than the line length balancing method. We believe that this final interpretation (Fig. 6-2d) is the most accurate interpretation with the data and models available. Observe that the hanging wall closure is still a solid prospect; however, the footwall play has been eliminated. Many well constrained folds do not contain upturned footwall beds (Suppe and Medwedeff 1990). The construction of turned up footwall beds is a common interpretation pitfall which has resulted in a number of dry holes. Strong evidence for upturned beds should exist before drilling these plays.

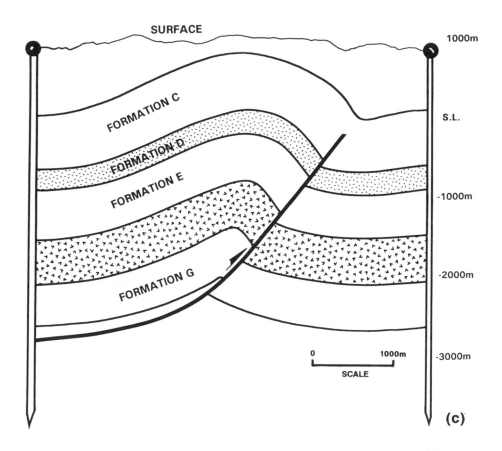

Figure 6-2c Reinterpretation of data using line length measurements. Bed length on top of Units D, E, and G are designed to conform to the top of shallow Unit C.

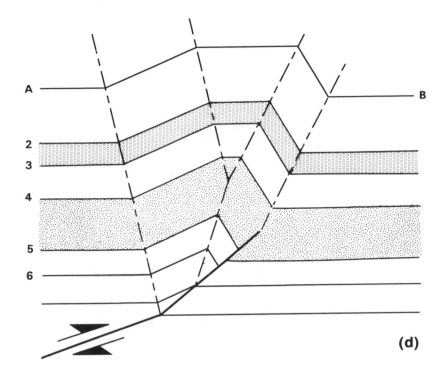

Figure 6-2d Reinterpretation of hanging wall data using the Kink method and the fault propagation fold model.

At times kink band folds are incorrectly interpreted as listric reverse faults. In duplexes (Figure 6-3a) kink band folding and deformed faults can be combined to provide a misinterpretation as a *continuously upward* listric reverse fault. Figure 6-3a illustrates the kink folds and kinked or deformed fault surfaces in a duplex. Figure 6-3b shows an incorrect, continuously upward listric reverse fault interpretation of the same cross section. In this forced interpretation, the fault follows an axial surface at shallower depths.

RETRODEFORMATION

Compressional or rollover anticlines do not form in place. Therefore, any interpretation must be retrodeformable or restorable to its initial or undeformed state. You may recognize this concept as palinspastic restoration or backstripping. Structures move from their initial undeformed position into their present position.

110

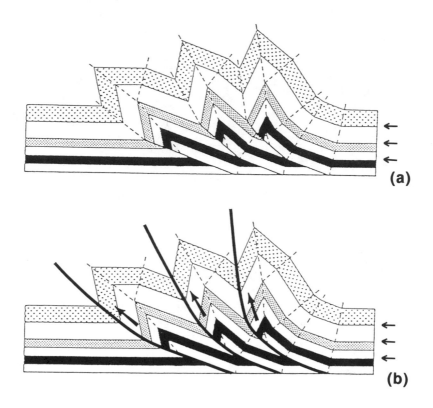

Figure 6-3 (a) Fault propagation fold duplex in which the lower *blind* thrust fault deforms the two shallower thrusts. The blind thrusts do not cut the shallow units. (b) Misinterpretation of blind faults drawn through high bed dips and along kinks or axial surfaces. (Published by permission of American Association of Petroleum Geologists.)

This restoration process can often be accomplished by using scissors and some tape to move structures back to a common stratigraphic level. The process can be applied to both compressional and extensional structures. For example, referring back to Figure 6-1a, take scissors and cut along the faults. Simply move the units back down the fault ramps and over the basement until the tops of Unit B are roughly on a *common level* (Fig. 6-1c). A horizontal line drawn on the level of the top of unit B will aid the restoration process. This follows as sedimentary units are initially deposited in a horizontal position. This line should also parallel the regional dip. Cut the fault blocks into small pieces to accomplish the restoration, taping them to another piece of paper.

The units are deposited without gaps or overlaps in the strata. Thus, the restoration process should fill any voids caused by moving the units back to a common level. Restoration is complicated by the fact that the units are folded during the deformation process. This means that straight lines are now curved. Curved lines can be straightened out by cutting the fault blocks into small pieces that are approximately straight.

Two major problems become evident with this exercise (Fig. 6-1c). First, Unit C is repeated or overlaps after restoration, which is not physically possible (see hachured horizon in Figure 6-1c). The hachured Layer C is probably miscorrelated as Unit B and not as Unit C, thus accounting for the uneven thinning of Unit B. Second, a large piece of puzzle remains to be inserted into a small. This suggests that one or more faults were misinterpreted. The restoration shown in Figure 6-1c suggests that the section is not balanced and that it contains several busts. An incorrect structural interpretation like this can lead to *expensive dry holes*.

THIN-SKINNED TECTONICS

Rich (1934) was the first to recognize that thrusting is "thin-skinned" or occurs along bedding surfaces that are commonly weak (perhaps overpressured) shale horizons. These bedding plane thrusts often ramp to higher structural levels and upon deformation, form folds (Fig. 6-4a and 6-4b) (Suppe 1983 and 1985; Boyer 1986; Tearpock and Bischke 1991). These ideas remained controversial until seismic reflection and deep crustal seismic profiling imaged thin-skinned structures. Although these ideas are now well known to interpreters, bedding plane thrusts are not always drawn on sections. Perhaps this is due to the fact that bedding parallel thrusts do not image on most dip sections. If a bedding plane thrust is not drawn on a section, the resulting interpretation may not conserve area. This could result in a dry hole or an unrecognized prospect. In the section on Quickly Recognizing Décollements later in this chapter, we show you a method for recognizing thrusts that run parallel to bedding. Bedding plane thrusts are located by recognizing where more steeply dipping beds directly overlie more gently dipping beds.

Thrust positioning is important for several reasons. First, thrust fault position controls the size and location of the structural trap. Second, thrust surfaces separate hanging wall beds from footwall beds. Thus, a question arises: are potential reservoir beds located in the hanging wall structural trap, or do they exist beneath the thrust in the footwall? Third, hydrocarbons may be trapped by the fault surface as in some Canadian fields. And finally, thrust geometry defines the maximum depth to look for additional prospects.

112

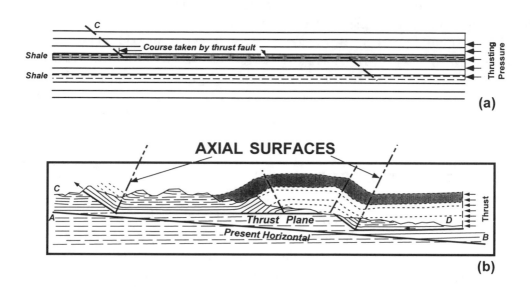

Course taken by thrust fault

Thrusting Pressure

Shale

Shale

(a)

AXIAL SURFACES

Thrust

Thrust Plane

Present Horizontal

(b)

Figure 6-4 (a) Pre-deformational ramp and flat geometry. (b) Slip of hanging wall block over footwall results in thin skinned tectonics and the generation of a symmetric or fault bend fold. Where beds move through fault bends, they form axial surface (Rich 1934. Published by permission of the American Association of Petroleum Geologists.)

An example of the latter reason is shown in Figure 6-5, which represents depth-corrected seismic reflections from an Asian fold-thrust belt. The bedding plane thrust fault is defined by the flat portions of the bold line. The upper level bedding plane thrust (left side of figure) defines the depth of prospect potential above the thrust. As the deformation exists on more than one structural level, additional ramps and their related anticlines may be present off the section at depths defined by d_1 and d_2 (compare to Fig. 6-4). For example, the dips located near the right hand side of Figure 6-5 may represent the front limb of another, deeper fold.

PROSPECTING LEVEL

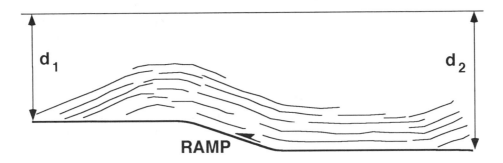

d₁ d_1 d₂ d_2 RAMP

Figure 6-5 Locating depths d_1 and d_2 at which to prospect for additional structures.

FOLD SHAPE IS RELATED TO FAULT SHAPE

Notice in Figure 6-4 that, if the hanging wall block slides over the footwall block, a fold forms. *The shape of the fault surface controls the shape and position of the fold.* The processes that form folds are beyond the scope of this book, but are described in detail in Suppe (1983, 1985), Mitra (1985), Boyer (1986), Suppe and Medwedeff (1990), Tearpock and Bischke (1991), and Suppe et al. (1992), and briefly in this book in the section on growth. If we know the shape of a fold, we can often predict the shape of the fault that formed the fold and, possibly, the position of the fault. Also notice that axial surfaces are generated where hanging wall beds move through fault bends (Fig. 6-4b). Where the beds move through bends in the fault surface, the beds **kink or bend** to accommodate the deformation. Therefore, axial surfaces can be used to help determine the limits of a prospect and where to look for additional structures. We discuss techniques for locating these important features in this chapter in the section entitled Dip Domain Analysis.

BALANCING AND THRUST FAULT STEP-UP ANGLES

In areas of complex thrust faulting and high bed dips, a structural interpretation should be balanced to insure cross section viability. Although structural balancing takes time, it is cheap relative to dry holes. If done correctly, balancing can often limit the requirement of having to "change the geology" every time a well is drilled. Balancing is most useful where data are poor or limited on seismic profiles and where structures are complex. This is often the case in fold belts. In this section, we present another QLT that will rapidly tell you whether the sections under review are in balance.

In highly deformed regions, thrust faults are not commonly imaged on record sections; however, thrust faults are often drawn through poor or no data areas of seismic sections, or through areas of high bed dip. Bed dips exceeding 25 to 30 deg may not image on conventional seismic sections. Several interpretation problems can arise due to the high bed dips. First, if the thrust fault is positioned too high, then a potential reservoir unit may be placed in the footwall (Fig. 6-6a). This would cause a well to be terminated at too shallow a depth, resulting in a potential hanging wall trap unit remaining unpenetrated. Second, if the poor data region located above the ramping thrust fault results from steep or overturned beds (no data zone in Fig. 6-6a), then this region is to be typically avoided during drilling. An example of this common pitfall and problems associated with steep bed dips are discussed in the next section. Available data indicate that only a limited number of structures produce exclusively from the front limb of anticlines. One such anticline is the Santicoy Oil Field, California, where production comes from updip stratigraphic pinch out and fault traps (Yeats and Taylor 1990).

One easy way to recognize these types of structural pitfalls is simply to notice if the thrust faults in a given area **step-up at a common angle** on existing cross sections (Fig. 6-6b). If a **listric** fault is drawn through the no data regions, then perhaps no attempt was made to distinguish hanging wall beds from footwall beds. Also, no attempt may have been made to distinguish between poor or no data zones and regions of high bed dip. This is a common, but costly, pitfall.

INCORRECT LISTRIC FAULT INTERPRETATION

FAULT TOO SHALLOW!

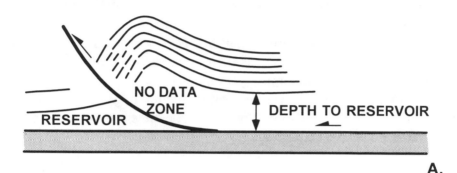

A.

CORRECT INTERPRETATION

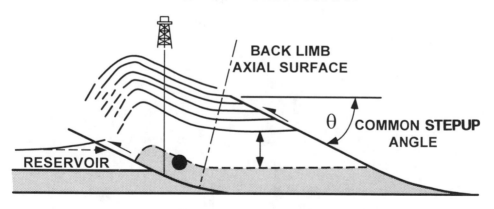

B.

Figure 6-6 (a) Listric thrust fault pitfall. This interpretation suggests that the reservoir unit is not involved in the folding. (b) Correct or balanced interpretation. Thrust faults in a given area typically step-up at a common angle, in this case including the reservoir unit in the folding (see Fig. 6-5).

Our experience in thrust belts throughout the world shows that thrusts have a tendency to step up at a **common or fundamental angle**. An example is shown in Figure 6-7, where thrusts in Taiwan step-up at angles of 9 to 17 deg or at 19 to 26 deg, depending upon whether the anticline is a symmetric or asymmetric structure, respectively. If the structure is stacked like a **duplex** (Suppe 1983; Mitra 1986 and 1988; Tearpock and Bischke 1991), which has the potential for **multiple objectives**, then the thrusts are most likely stacked in rough increments of the common step-up angle. For example, if the common or fundamental step-up angle in an area is 20 deg, then volume conservation requires that thrusts on a structurally higher level dip at about 38 deg, or slightly less than twice the common step-up angle. In this example, the 38 deg dipping thrust or beds in a structural stack are unlikely to image on all but the best migrated data. We must note here, however, that regional conditions and rock properties are known to produce changes in the common step-up angle.

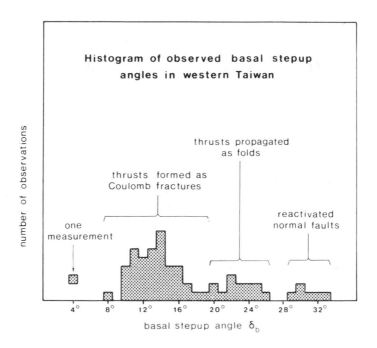

Figure 6-7 In Taiwan, symmetric or fault bend folds step-up at 9 to 16 deg angles, while asymmetric or fault propagation folds step-up at between 20 to 26 deg angles. (From Dahlen, Davis and Suppe, 1984, *Mechanics of Fold and Thrust Belts and Accretionary Wedges*: Jour of Geophysics Res, v. 89, p. 10087-10101, copyright by the American Geophysical Union.)

DO THE AXIAL SURFACES BISECT THE FOLD LIMBS-OR OFF STRUCTURE AGAIN?

In this section, we discuss one of the most common dry hole pitfalls encountered in compressional structural geology, the **Off Structure** problem (Bischke 1994). Because of its importance, we devote more discussion to this topic. In compressional (but not in extensional) terrains, the continuity of the bed length and thickness rules generally requires that the axial surfaces **bisect the fold limbs** (Suppe 1985; Tearpock and Bischke 1991). Although some **asymmetric** structures exist in which the limbs of the folds do not bisect (Mitra 1990), most structures deviate only slightly from the **Fold Limb Bisecting Rule** (Suppe and Medwedeff 1990). The extent to which asymmetric structures depart from this rule is shown in Figure 6-8. This figure is a collection of some of the best constrained examples of large (economic scale) asymmetric structures in the world.

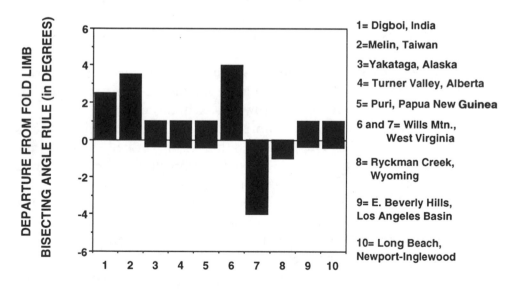

Figure 6-8 Dip data taken from well constrained petroleum size structures from throughout the world indicate that the "Fold Limb Bisecting Rule" roughly applies to most asymmetric structures. The graph explains why wells drilled on top of asymmetric folds typically encounter high bed dips throughout the well without penetrating the reservoir horizon.

The structures in Figure 6-8 are constrained by good outcrop exposure and good surface dips, or by well and dipmeter data. For well constrained examples, the dipmeter data must penetrate both the front and back limbs of the structure.

Figure 6-8 shows that most asymmetric structures tend to deviate less than 5 deg from the Fold Limb Bisecting Rule. For example, if the angle between the back and front limbs of a fold is 90 deg, then the bisecting rule requires that the axial surface is located 45 deg from both limbs. Figure 6-8 requires that the axial surface will lie within 45 deg $\pm$5 deg from both limbs. Thus, an axial surface dipping at 40 deg relative to the back limb and 50 deg to the front limb is possible.

Unfortunately, geologic cross sections through prospects are often generated at exaggerated scales. Sections that are not generated at true scale distort geometry and do not conserve volume or area (Tearpock and Bischke 1991). It is not easy to recognize whether or not axial surfaces bisect fold limbs on seismic data or on cross sections that are *not* generated on a scale of one-to-one. Thus, workers should be encouraged to construct cross sections for prospects at a scale of one-to-one, and managers should insist on true scale. **Distorting the true geometry of a prospect is a major structural pitfall.** Always remember that seismic sections distort geometry.

If the limbs of compressional folds are not approximately bisected by their axial surfaces, then one does not need a well to know that "the geology will probably be changed" after the well is drilled. This rule only applies to *pre-growth sediments* or to sediments which are *not syntectonic*. The rule applies to beds that do not rapidly change thickness. This topic will be discussed further in the section on Growth later in this chapter.

A cross section of an asymmetric or fault propagation fold is presented in Figure 6-9a. This interpretation was generated using seismic data, surface geology, and surface dip data. Unfortunately most seismic profiles are horizontally compressed, which may mislead interpreters to the interpretation shown in Figure 6-9a. Observe that the axial surface in the core of the fold does not bisect the front and back limb dips ($\theta_1 > \theta_2$). A closer examination of the interpreted fold reveals that the front limb of the structure thins relative to the back limb.

There seems to be a prejudice among some geoscientists to assume that the front limbs of folds thin. This prejudice may originate from structural geology texts written from a high temperature or metamorphic structural geology point of view. Indeed, most textbooks on structural geology have this bias. However, work on well constrained large scale folds

A MAJOR PITFALL,
FOLD LIMBS DO NOT BISECT

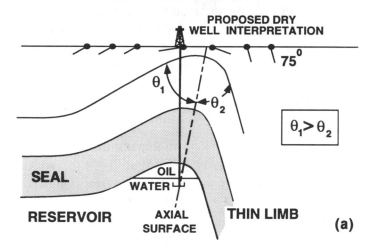

INCORRECT INTERPRETATION

PROPOSED DRY
WELL INTERPRETATION

75^0

θ_1

θ_2

$\theta_1 > \theta_2$

SEAL

OIL

WATER

RESERVOIR

AXIAL
SURFACE

THIN LIMB

(a)

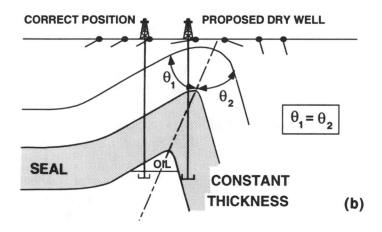

CORRECT INTERPRETATION

CORRECT POSITION

PROPOSED DRY WELL

θ_1

θ_2

$\theta_1 = \theta_2$

SEAL

OIL

**CONSTANT
THICKNESS**

(b)

Figure 6-9 Not bisecting fold limbs, a common structural pitfall. (a) The limbs of the fold on this cross section do not bisect, suggesting that the front limb thins dramatically. This assumption commonly results in positioning the well on top of the structure and results in a dry hole. (b) Correct or balanced interpretation shows that well should be spudded near the back limb of the structure, often where the beds start to increase dip.

(Suppe 1983 and 1985; Suppe and Medwedeff 1990) (Fig. 6-8) and on outcrop scale structures (Mitra 1990) situated within the lower temperature portions of fold belts demonstrates that folds possessing thicker frontal limbs may be more common than folds possessing thinner front limbs. Many large scale structures contain front limbs that are folded, suggesting compression and thickening of the front limb. The well-known Turner Valley Anticline is this type of structure (Suppe and Medwedeff 1990).

Proper Well Position

Let us closely examine the consequences of an interpretation that assumes the front limb of a fold thins. In Figure 6-9a, consider the lowest unit, which is labeled *"reservoir"*, to be the target unit. In this example Θ_1 = 50 deg and Θ_2 = 27 degrees. On Figure 6-8, this corresponds to a departure angle of (50 deg + 27 deg)/2.0 - 50 deg = -11.5 degrees. The large negative number predicts that this fold has a thinner front limb and thus plots off scale on Figure 6-8.

This more extreme geometry may not commonly exist within the lower temperature portions of fold belts. A well, positioned near the crest of the structure will intersect the 75 deg dipping beds within the stippled seal horizon (Fig. 6-9b). This occurs because most asymmetric folds tend to be more asymmetric than shown in Figure 6-9a. Figure 6-9b is a common geometry encountered within fold belts (Tearpock and Bischke 1991). It is, however, not the only geometry encountered. This interpretation maintains bed length and thickness consistency, which requires that $\Theta_1 \approx \Theta_2$ (Fig. 6-8). An interpretation generated to encounter a productive horizon in an optimum position requires that the well be spudded on the back limb of this asymmetric structure, commonly called a *snake head or fault propagation fold* (Suppe 1985; Tearpock and Bischke 1991).

If the structure has a front limb that thickens, then the structure will be even more asymmetric than shown in Figure 6-9b. Figure 6-9c illustrates an interpreted fold with a 12.5 deg departure angle, which may be at the extreme limits of the limb thickening type asymmetric folds. In these fold types, the reservoir crest is shifted even further from the top of the shallow surface structure. Notice, however, if we assume that the limbs of the fold bisect as shown in Figure 6-9b, we may still make the discovery (Fig. 6-9c). We must stress that **extreme care** should be exercised when positioning wells on asymmetric structures.

Our study of cross sections through asymmetric structures around the world strongly suggests that many wells drilled off-structure are the result of not correctly bisecting fold limbs. This pitfall also applies to symmetric or fault bend fold styles shown in Figures 6-4 and 6-5. Incorrect interpretations can cause wells proposed on top of the structure to actually be drilled in the steeply dipping front limb beds (Bischke 1994). Furthermore, constructing cross sections with a **vertical exaggeration** distorts geometry and can add to the problem (Tearpock and Bischke 1991). Seismic sections are commonly vertically exaggerated and also distort true geometry. These are simple, yet common and costly, structural pitfalls.

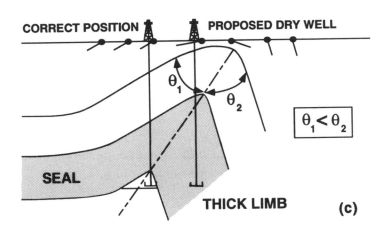

Figure 6-9c If the structure contains a folded or thicker front limb, then the fold may be more asymmetric than depicted in Figure 6-9b. These fold types plot as positive values on Figure 6-8 and may be more common than thin limb folds, which plot as negative values on Figure 6-8.

Let's review an example through three stages of structural evolution. Figure 6-10a is a predrill interpretation of an asymmetric fold in which the field geologists did not recognize that a massive carbonate unit (Formation U) contained an overturned frontal limb. The interpretation is primarily based on outcrop and surface dip data. The interpretation indicates that the proposed well will penetrate the crest or back limb of the structure.

122

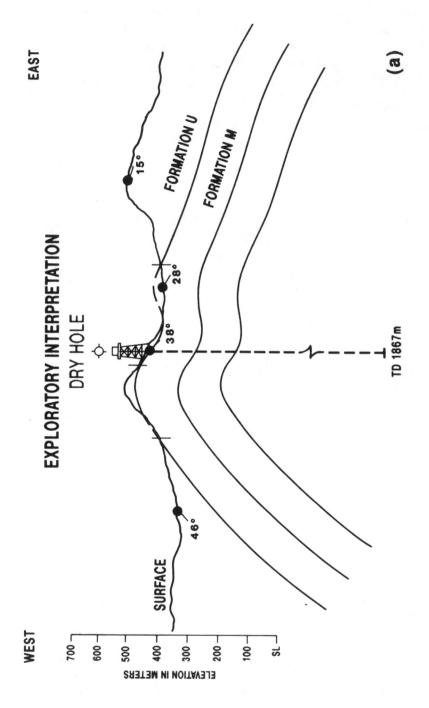

Figure 6-10a Surface outcrop and dip data suggest the presence of a broad anticline located in the Philippines. (Published by permission of Philippine National Oil Company.)

123

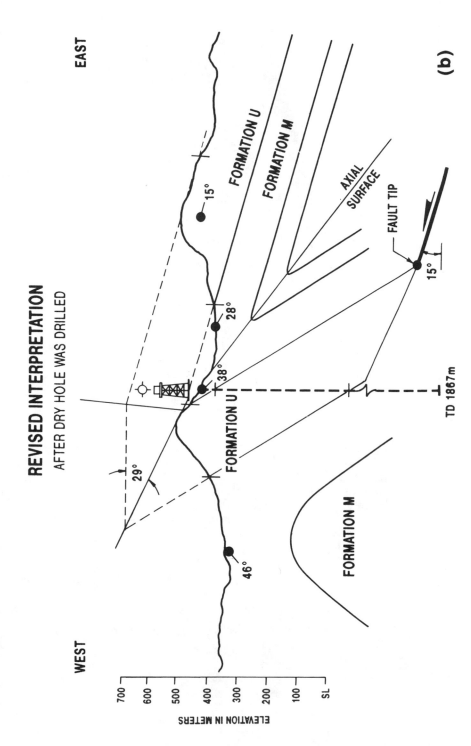

Figure 6-10b Reinterpretation of original data, incorporating well log, additional field work, and the Kink method. (Published by permission of Philippine National Oil Company.)

124

RAM FOLDS

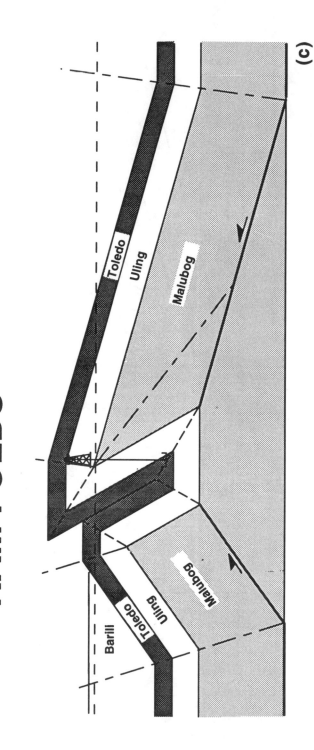

Figure 6-10c Generic interpretation of structures present in Figure 6-10b. (Published by permission of Philippine National Oil Company.)

After a dry hole was drilled that encountered a thick section of carbonates, the structure was reinterpreted (Fig. 6-10b) using dip domain construction and fault propagation fold modeling. The new interpretation resulted in a major change in the structural picture. With the new interpretation, it can be seen why the well actually penetrated the nonproductive, steep, overturned, front limb rather then the crest or back limb of the structure as initially interpreted.

Figure 6-10c is a final generic interpretation of the structure that contains an unconformity and a period of non-deposition at the top of the Toledo Formation. It is interpreted as a Ram structure similar to the interpretation of the Painter Reservoir Field, Wyoming. We have seen a number of such examples in fold and thrust belts around the world.

THRUST FAULT POSITION

Another technique that can be used to check structural balance rapidly and to determine prospect size and reservoir position is to observe if the back synclinal limb of the fold is bisected and if the shape of the fault corresponds to the shape of the fold. Returning to unbalanced Figure 6-6a, we see that no relationship exists between the shape of the fold and the shape of the fault. Observe that the back limb beds dip at a **constant angle** above a fault that **curves** (Fig. 6-6a). If hanging wall beds move up a curved fault, then the hanging wall beds must curve above the curved fault to accommodate the volume of rock moving up the ramp. **Thus curved faults produce curved folds**. Interpretations of the type shown in Figure 6-6a do not conserve area and, typically, remove volume from the core of the fold. However, notice in the balanced section shown in Figure 6-6b that the **fold shape mimics the fault shape.** In other words, the shape of the beds in Figure 6-6b conforms to the shape of the fault.

The structural problem encountered in Figure 6-6a can be solved by locating the approximate position of the thrust fault on the seismic section. In asymmetric structures, the hanging wall beds at the front of the structure typically dip at high angles. The thrust at the front of the structure is positioned at the intersection of the steeply dipping beds with the deeper, more gently dipping footwall beds as shown in Figure 6-6b (Suppe 1985). The footwall beds are normally imaged on seismic sections, while hanging wall beds are represented by no data zones. Therefore, the thrust fault can probably be positioned along a line that is defined by the termination of the footwall beds, or where the data deteriorate as a result of high bed dips in the hanging wall. Remember that the region below the fault is subject to pullup, which is accounted for in Figure 6-6b (see dashed arrow). By applying these basic principles, the approximate position of the thrust can be located and projected forward, conforming to hanging wall bed dips.

Thrust faults will usually dip at the **same or at a higher angle** than the hanging wall bed dips (Suppe and Medwedeff 1990). If we apply the Kink method to this fold (Fig. 6-6b), we can bisect the fold's back limb beds, defining the synclinal axial surface. We can now project the thrust fault down to this axial surface, parallel to the hanging wall bed dips. Since optical illusions abound, you should align a parallel glider or straight edge parallel to the bed dips, and then slide the glider down to the position of the fault pick. The glider can now be used to project the fault forward. If the bed dips curve, then the fault is listric and will follow a curved line that parallels the bed dips. The fault will flatten out at the axial surface in the region of the syncline. The fault then subparallels the flat bed dips located at the back of the structure (Figs. 6-1b, 6-4b, 6-5, and 6-6b). Using these geometric techniques you can rapidly determine the shape of the fault and fold. The position of the fault then allows you to determine if the objective beds exist in the hanging wall.

Dip Domain Analysis:
Rapidly Locating Thrusts and Plays

In previous sections, we described the relationship between fault shape and fold shape, and that bends in fault surfaces generate folds with axial surfaces. The axial surfaces which emanate from fault bends cause the beds to change dip. Thus, **axial surfaces can be used to interpret structure and to locate fault bends and position.** Fault shape and position are important to any evaluation, as they control the existence and the size of a prospect.

In Figure 6-11a, notice that broad regions of near constant dip are separated from narrow regions of rapidly changing dip. The regions of rapidly changing dip on this figure define the **axial surfaces,** which are shown as the more steeply dipping thin lines. Those regions that contain coherent reflectors of near constant dip are called **dip domains or dip panels.** Each dip domain is separated from an adjoining dip domain by axial surfaces or by faults. Dip domain analysis has the advantage of representing an objective concept, based upon uniformly dipping packages of coherent reflections. Therefore, *dip domain analysis is less interpretative*. Interpreters who use dip domain analysis are thus more *objective* in their interpretations.

Dip domains are best isolated by aligning a parallel glider or rolling ruler parallel to packages of coherent seismic reflections, passing the ruler across the packages of uniformly dipping beds, and then marking on the section the positions where the beds change dip (consult Fig. 7-9, in Chapter 7). Every significant change in bed dip results in the introduction of an additional dip domain. The position of every dip domain should be confirmed in the adjacent dip panel. This is accomplished by passing the rolling ruler or parallel glider across the adjacent data. After a little practice, this technique allows one to distinguish axial surfaces from faults and accurately determine where faults are located. In pre-growth sediments, the axial surfaces bisect the bed dips or dip panels. This principle has also been used to correctly calculate interval velocities on seismic sections (Bischke 1990).

After isolating the individual dip domains, you can interpret the seismic section. Faults by definition, lie along discontinuities or offsets in the dip domains (bold lines in Figure 6-11b) or along bedding surfaces. An example of a large discontinuity in bed dips is shown in Figure 6-11a at sp A (at 15,000 to 25,000 ft), where dipping reflectors overlie flat reflectors. You must also remember that faults should not cut coherent packages of reflectors that do not change reflection character. *In areas of complex deformation, drawing faults through packages of coherent reflectors is a common pitfall.* For example, faults do not lie along axial surfaces, or where the beds kink or uniformly change dip. An example of an axial surface is shown in Figure 6-11a at sp B (5,000 to 20,000 ft). Here, coherent reflectors that do not change reflection character are observed to cross the axial surface without being offset.

By honoring dip panels, dip domain analysis forces interpreters to position thrust faults more accurately and objectively. This analysis has the advantage of *honoring the data prior to interpretation,* so interpreters are less likely to draw in faults where faults are not present. The interpretation of the dip domains is shown in Figure 6-11a. Faults shown in Figure 6-11b are mapped by following discontinuities at the edges of the panels or along bedding surfaces, as bedding plane thrusts.

128

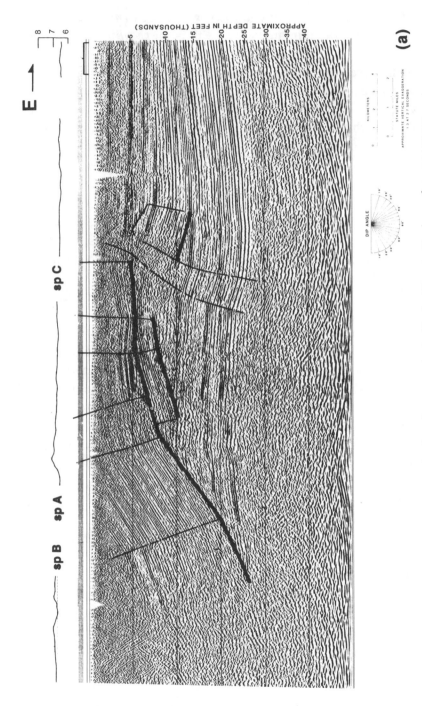

Figure 6-11a Seismic section imaging ramps in the Darby Thrust. Regions of near constant dip (dip domain) are separated from other regions of near constant dip by axial surfaces. (See Williams and Dixon (1985). From Seismic Exploration of Rocky Mountains Region. Seismic published by permission of the Rocky Mountain Association of Geologists and Denver Geophysical Society. Interpretation by Richard Bischke.)

129

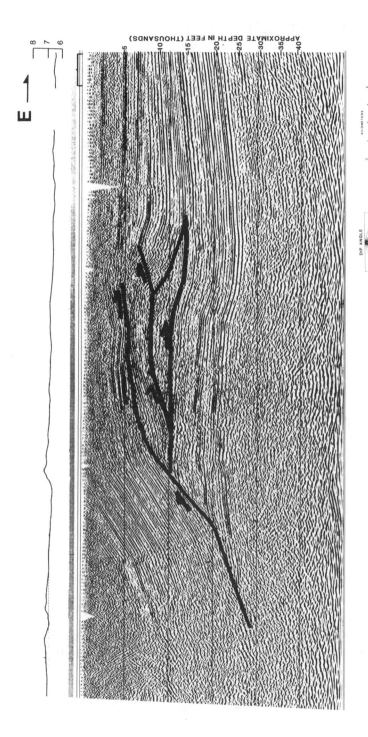

(b)

Figure 6-11b Interpreted seismic section. Notice that thrust faults exist between dip domains, generally where more highly dipping beds overlie more gently dipping beds. Faults can also follow bedding unconformities. (See Williams and Dixon (1985). From Seismic Exploration of Rocky Mountains Region. Seismic published by permission of the Rocky Mountain Association of Geologists and Denver Geophysical Society. Interpretation by Richard Bischke.)

Quickly Recognizing Décollements

In the roughly balanced section shown in Figure 6-1b, a décollement can be recognized by noticing that the bed length at the base of Unit C is much shorter than bed lengths along the tops of Units C or B. This is Dahlstrom's (1969) classic method for detecting major detachments (Tearpock and Bischke 1991).

Perhaps a quicker method for recognizing major detachments or thrusts is to *locate areas where more steeply dipping beds* overlie flatter beds (Figs. 6-1b or 6-11b). In Figure 6-1b, dipping beds in the core of the anticline overlie more gently dipping footwall beds. As shown previously, an example of a large change in bed dip, where highly dipping beds overlie flat beds, is imaged on the seismic profile shown in Figure 6-11a, at sp A between 15,000 and 25,000 ft (below 5,000 meters). The Darby Thrust ramps upward in this region to a higher structural level, and the discontinuity between dipping and flatter beds identifies this major thrust fault surface.

Interpreters active in normal fault terrains also use this criterion, perhaps subconsciously. In extensional areas, the discontinuity between the hanging wall and footwall beds is often well imaged on seismic sections. Mappers may not notice that they are also applying the change in bed dip criterion when mapping major normal fault surfaces.

Of course, changes in dip caused by sedimentary processes, such as baselap, are an exception to this rule. If, however, the *change in bed dip* is larger than that for typical sedimentary bed dip (commonly less than 5 deg, Rich 1951), then structural deformation must be considered.

Rapidly Locating Additional Prospects

In the last section, we used the concept of dip domain panels to map thrust faults on seismic sections. This concept can be used to locate deeper or more poorly imaged prospects. In Figure 6-11a, a deeper dip panel exists at sp C beneath the thrust located at the 13,000 to 15,000-ft level. This deeper panel rolls over beds between the 13,000 to 25,000-ft (4000 to 7500 meters) level. Detailed mapping would have located this deeper structure, but dip panel analysis allows one to **quickly** recognize the deeper structure.

A KINK OR A FAULT?

So far we have learned that one method for identifying the presence of **thrust faults** is to locate **dipping beds over flatter beds,** which indicate a discontinuity in dip on a seismic section. However, faults do not always exist where the beds simply change dip or where there is an apparent discontinuity. You must question the cause of the discontinuity or the change in bed dip. Changes in bed dip or apparent discontinuities on seismic sections can result from several causes and **not just faulting.** Among these are:

1. improper migration, particularly in areas of high bed dip;
2. poor or incorrect interval velocities, or abrupt horizontal velocity changes;
3. velocity pull-up or sag;
4. abrupt dip changes **(kinks)** of the beds, which occur at **axial surfaces;** and
5. angular unconformities.

Of course, drawing in a fault where a fault is not present can create a non-existent prospect or destroy a good prospect. Additionally, incorrect fault interpretations can result in incorrect regional structural interpretations, as well as incorrect reservoir maps, where reservoirs are broken up by numerous faults.

An example of abrupt changes in dip caused by high bed dips is shown in Figure 6-12, from the Canadian Rockies. This kink or chevron fold style is common in thrust belts where interbedded sedimentary sequences have been folded. Additional examples are presented in Suppe (1985), Woodard et al. (1985), Boyer (1986), and Tearpock and Bischke (1991). If a seismic line was shot across this kink fold, the high bed dips would not image on the section. An interpreter might conclude that a fault zone exists within the high bed dip region, located between the anticlinal and synclinal axial surfaces.

A similar, but less dramatic, example is shown on the well imaged seismic section in Figure 6-13. This section images a flat topped anticline (box fold) with two rapidly changing dip domain panels on both limbs of the structure. If faults are drawn through these areas of abrupt change in dip, then *a prospect may be destroyed or non-existent fault traps created.* If, in the high dip areas, improper migration and side-swipe are taken into consideration, then there is little or no evidence that faults offset the reflections within the core of this structure (Fig. 6-13). This can be

132

confirmed by duplicating the figure and correlating the limb reflections into the core of the fold. After this is done, we see that deterioration of the seismic data and apparent discontinuities can result from changing bed dips, caused by a **kinking** of the beds.

Figure 6-12 Outcrop of kink or Chevron type fold, Canadian Rockies. Regions of shallower bed dips are separated from regions of higher bed dips by axial surfaces. High bed dip regions will not image on seismic sections and may be interpreted as a zone of faulting. (From Journal of Structural Geology, v. 8, Boyer, *Styles of folding within thrust sheets, examples from Appalachian and Rocky Mountains of the USA and Canada*, 1986. Pergamon Press Ltd.)

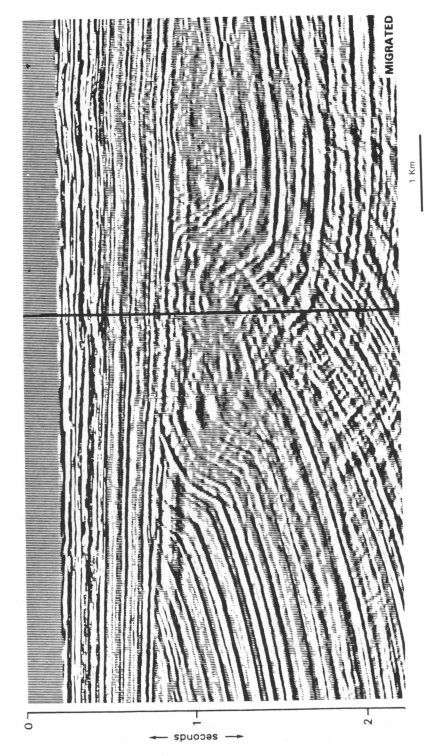

Figure 6-13 Areas of rapidly changing dip on seismic sections are sometimes interpreted to be fault surfaces. This can be checked by correlating reflections across the interpreted "fault surface", and then examining the section for offset reflectors. (From Goudsward and Jenyon 1988. Published by permission of the European Association of Exploration Geophysicists.)

We also know that high bed dips often result in improper **migration,** and that high dips will cause the reflections to change character across axial surfaces. The general approach to this problem is to ask, **"If the fold were not present, would I draw a fault through areas that do not contain offset reflectors"?** Of course you would not. The existence of the fold does not change this basic principle. The presence of faulting can often be checked by correlating the reflectors across the "faulted surface". If there is no offset in the reflectors, then no fault is present. This discussion may seem elementary, yet, in highly folded areas, it represents a common structural pitfall. This pitfall is particularly common in areas where wrench faulting and folding are believed to exist.

Returning to Figure 6-13, we conclude that the beds are folded across axial surfaces that separate the fold limbs from its crest, and that the changing bed dips *are not* caused by faults, but by axial surfaces, or **kinks.** Thus, the structure can be interpreted to be a complex symmetric fold or box like structure.

FAULTS OR A BOX FOLD?

Several interpretation problems occur in a class of structures called box or lift-off folds. These structural folds are now known to exist in the North American Rockies, Switzerland, South America, Asia, and elsewhere (Laubscher 1977; Namson 1981). The structures typically form over very weak incompetent horizons such as evaporites, or thick overpressured shale intervals that tend to flow.

Box and lift-off structures possess steeply dipping to vertical limbs which may not image on seismic sections. There is a strong temptation for all of us to *draw fault zones through no data regions,* thus solving the no data problem - or so we think. Poor data can, however, result from a number of factors including:

1. highly dipping beds,
2. a host of data acquisition problems,
3. folding, or
4. faulting.

Field acquisition parameters designed for high dip areas, pre-stack migration, and velocity modeling are often used to image high dips.

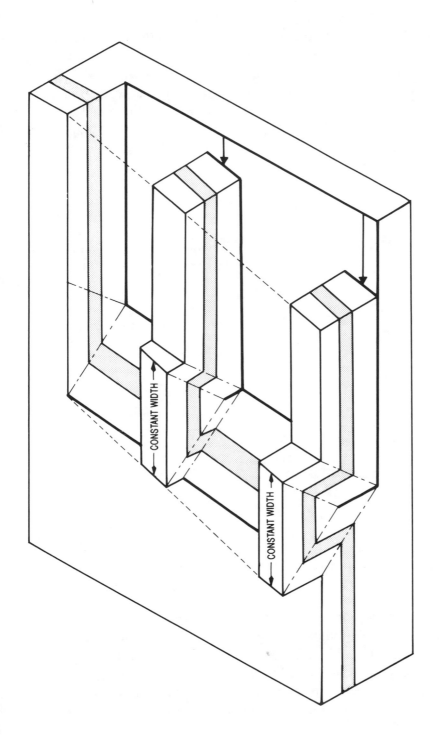

CONSTANT WIDTH

CONSTANT WIDTH

Figure 6-14 In box folding, the width of the structure across its crest remains about constant along strike. Notice that the folding causes the limbs on box folds to dip at near vertical angles, and, thus, the upturned beds will not image on seismic sections.

Box folds have this particular characteristic: the *width* of the structure across its crest (top) remains about *constant along strike* (Fig. 6-14). If a map is drawn showing two straight, parallel wrench fault zones that are separated from each other by a region of nearly constant width (Fig. 6-15), one should consider the presence of box folding rather than wrench faulting. **Wrench faults,** like box folds, can turn up strata, and then tend to occur along straight parallel trends (Shaw et al. 1994). These two fundamentally different structural styles can occur in the same geologic province and can be mistaken for each other.

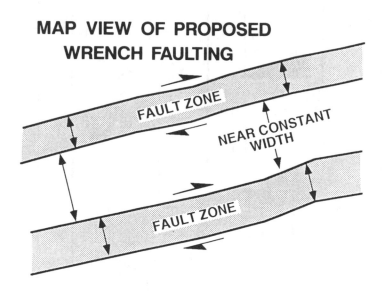

MAP VIEW OF PROPOSED WRENCH FAULTING

Figure 6-15 Two proposed wrench fault zones that parallel each other at near constant width could represent vertical beds caused by box folding (see Fig. 6-14). Other interpretations are also possible.

In good data regions, one way of distinguishing between the two structures is to notice that in box folding, the reflection character at the crest of the fold will correlate with reflections on both flanks that are on roughly the same, yet deeper, structural level. For example, in Figure 6-13, reflections on both flanks of this structure can be correlated onto the crest of the structure. If this can be done, we can question the presence of wrench faulting. If faulting is indeed present, then displacements along these faults are probably not large.

LIMITED STRUCTURAL STYLES

A prospect can be quickly rejected if strange structural relationships are noticed on the interpreted cross sections or seismic sections. Dahlstrom (1969) recognized that a limited number of structural styles are common within a given structural environment. Some working experience is required in an area before you can recognize which styles or structures are present. However, one should always keep an open mind for the unexpected.

We have observed, on seismic lines from around the world, good examples of normal faulting in fold belts, that have occurred both before and after compression. One common method for identifying precompressive normal faulting is to recognize **inversion structures** on seismic sections. In these structures, the folded beds are often thicker in the core of the fold and thin off structure. Structurally higher beds may be seen to thin onto the fold crest. In addition, graben structures can form at the crest of growing anticlines and therefore, are contemporaneous with the regional compression. An example of this structural style is shown in Figure 6-1b.

Extensional terrains can contain compressional structures, often in the deep water toe region of the major extensional system. In one extensional terrain, using 3-D seismic data, we have observed folds that were apparently driven by the lateral motion of salt sills. On conventional 2-D lines, these structures were initially interpreted to be salt domes, but they contained no salt when penetrated by drilling. When reviewed with 3-D seismic data, the structures turned out to be asymmetric compressional folds formed by high angle reverse faults.

A geologist must rely on experience and knowledge of an area to guide interpretations. Therefore, one would not expect to observe thrust faults on the flank of a growing salt dome. Repeated section on the flanks of salt domes can occur during normal faulting, if the dip of the *upturned beds exceeds normal fault dips* (Tearpock and Bischke 1991). In this case, repeated section should not be misinterpreted as a reverse fault. Although exceptions to the limited structural styles rule exist, common sense and questioning of the uncommon is *always* necessary where investment dollars are concerned.

Briefly, we list some common faults and associated structures found in particular structural settings.

A.   Compressional

1.   Symmetric and asymmetric folds, such as fault bend folds, box folds, and fault propagation folds. Ramp and flat thrust faults, imbricate thrusts, wedges, triangle zones, and "Rabbit Ears" (Mitra 1986; Suppe 1985; Tearpock and Bischke 1991.)

2.   Compressional duplexes.

3.   Shale diapirs.

4.   Lateral transfer zones.

5.   Regional wrench faulting is sometimes, but not always, present.

B.   Extensional

1.   Rollover structures formed by large listric normal fault systems. Listric and antilistric (concave down) shaped faults, several types of intersecting faults, including compensating and bifurcating patterns, and downward dying growth faults.

2.   Extensional duplexes.

3.   Shale or salt diapirs.

C.   Wrench

1.   Folds at constraining bends in wrench faults and rollovers at releasing bends (Crowell, 1974).

2.   Often wrench faults and associated folds exist in the same area (Harding 1990; Wright 1991), but may be independent of each other .

MAP AND CROSS SECTION CONSISTENCY

Dahlstrom (1969) noticed that interpretations on two adjacent seismic lines or cross sectional profiles must be **consistent**. In other words, large folds or thrust faults do not abruptly change or vanish from maps or adjacent cross sections. If this is observed, then related structures must exist to accommodate the abrupt change.

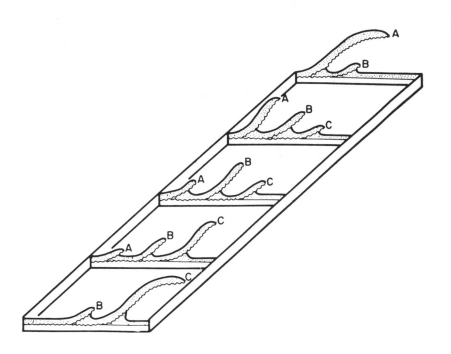

Figure 6-16 Slip transfer between Faults A, B, and C. From top to bottom, slip on dying Fault A is transferred to Faults B and C, etc., and, thus, Fault A is replaced by Fault C. (Reproduced by permission of the National Research Council of Canada, from the Canadian Journal of Earth Sciences, v. 6, p. 743-757, 1969.)

Dahlstrom illustrated this consistency principle (see Figure 6-16). There are two folds and thrusts on the uppermost profile (Folds A and B). If Fold A is smaller on an adjoining profile, then fault slip consistency typically requires that the deformation be accommodated by growth on Fold B, by the introduction of Fold C, or both. On the adjacent profile, slip on dying thrust Fault A is accommodated by growth on Faults B and C, etc. (Fig. 6-16). These relationships are commonly observed in fold belts and are responsible for the often observed relationship in map view of one fold dying and plunging to depth, while an adjacent or nearby fold terminates and plunges to depth in the *opposite direction* (Tearpock and Bischke 1991). This is caused by a uniform deformational or displacement **transfer zone** across the fold belt. This uniform transfer of displacement is a consistency principle that must be honored on seismic sections, cross sections, and maps. If deformation terminates **abruptly** along the strike of a structure and is *not accommodated* by another structure, cross structure, or fault, then the interpretation should be questioned.

BOW and ARROW RULE

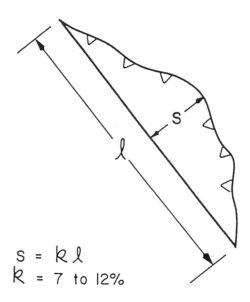

$$S = k\ell$$
$$k = 7 \text{ to } 12\%$$

Figure 6-17 Bow and Arrow Rule. Slip on thrust is roughly equal to 10% of thrust length. (From Elliot 1976. Published by permission of the Royal Society of London.)

You may be questioning this uniform displacement principle, since thrust faults are known to grow and die out along strike and in the core of asymmetric folds. Elliot (1976) developed the **Bow and Arrow** Rule to account for the more uniform changes in slip observed along fold-thrust belts. This empirical rule states that the maximum displacement or slip (S) along a thrust fault is 10% ± 3% of its length (L) (Fig. 6-17), or:

$$S = (0.1 \pm 0.03) \times L \qquad \text{Equation 6-1}$$

In the case of asymmetric folds, the slip along the thrust is not large and is accommodated by deformation and bedding plane slip within the core of the anticline (Suppe and Medwedeff 1990; Mitra 1990). The bedding plane slip consumes the slip on the fault and causes the fault to die abruptly.

CHECKING SLIP FOR CROSS-SECTION CONSISTENCY

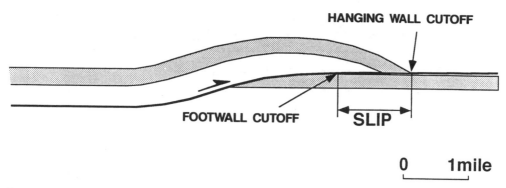

Figure 6-18 Checking for slip consistency. Amount of slip, defined by the distance between hanging wall and footwall cutoffs, should be about 10% of thrust length. Figure assumes that only one structure is present.

The Bow and Arrow Rule can be used as a QLT to check the size of a fault generated prospect. For example, in Figure 6-18, assume that you are working on a thrust that is known to gradually die out 5 mi from your prospect area. This means that the thrust must be at least 10-mi long. You can now measure on a cross section the amount of slip or displacement on a unit that is offset during thrust faulting (Fig. 6-18). Choose a unit and measure (with a ruler) the distance that the unit has been displaced between the footwall and the hanging wall cutoffs. In Figure 6-18, using the top of the stippled unit to measure the slip, we find that the slip is 5,500 ft. According to Eq. 6-1 and knowing that the thrust is at least 10-mi long, the amount of slip on the unit should be at least 1 ± 0.3 miles, which is within the range of the Bow and Arrow Rule. It appears from this QLT that the interpretation is reasonable.

DUPLEX STRUCTURES:
HOW TO RECOGNIZE MULTIPLE PLAYS

Duplex structures are as important as they are difficult. They present the possibility of **plays on more than one structural level.** These structures come in many forms and styles. They are produced by the stacking or repetition of a stratigraphic unit or units (Mitra 1986), as shown in Figure 6-19. The structures can, in a general sense, be recognized or verified by asking one or more of the following QLT questions.

142

1.   Do the wellbore correlations show a stratigraphic horizon repeated at least once?

2.   Does the seismic section indicate steeply dipping beds lying over more gently dipping beds, over still flatter dipping beds (Mitra 1986 and 1988)?

3.   Is the overall fold shape more rounded than expected?

4.   How many different dip domains exist on the front or back limb of a shallow fold?

Remember, these are quick look techniques, that in most cases should be followed up by more detailed analysis.

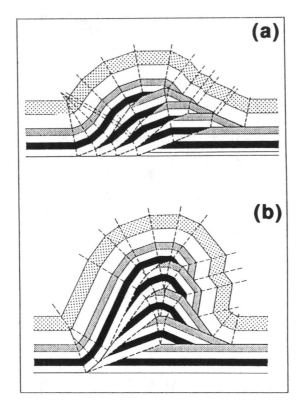

Figure 6-19 Stacked duplex structure, (a) for a small amount of slip. Multiple thrusts result in multiple plays. (b) Duplex amplified if slip is increased. (From Mitra 1986. Reprinted by permission of the American Association of Petroleum Geologists.)

Figure 6-19, taken from Mitra (1986), illustrates this more rounded fold shape, along with a black colored horizon that could represent repeated reservoir units. We should emphasize, however, that there is often a bias to make *folds more rounded on cross sections than the data warrant*. Furthermore, on seismic sections, Freznell zone interference causes folds to appear more rounded than they actually are in nature. Correctly migrating and depth correcting seismic data normally solves these problems and is certainly worth the cost to assure a successful exploration effort.

To further help recognize whether a duplex structure is present, dip domain analysis can be conducted on the back limb of the structure. If more than one dip domain or bed dip is present, then multiple thrusts may be present (Fig. 6-19). Now that the stacking by multiple thrusts is recognized, is there any quick method of determining the number of stacked thrusts and, therefore, repeated sections, which may prove to be prospective? Although this is a somewhat complex subject (Mitra 1986; Suppe 1983; Tearpock and Bischke 1991), there is a simple rule of thumb that can be applied. *Using the regional dip as a frame of reference, count the number of different back limb dip domains. This number is less than or equal to the number of stacked thrusts and potential repeated sections in the duplex.* In Mitra's generic example (Fig. 6-19b), there are four *different* back limb dips, and, thus, four potential repeated sections. Determining the number of repeated sections is best obtained from *properly collected* surface dip, subsurface dipmeter, and good depth corrected seismic data. These data define the individual dip domains. Isolate the number of dip domains by generalizing (i.e., averaging) the dip data. Typically, dips within each domain do not vary by more than a few degrees. Taking dips from the back or more gently dipping limb of the structure simplifies the dip domain analysis. The bed dips within each domain will occur in rough increments of slightly less than twice, three times etc., the lowest bed dip. The regional or 0 deg bed dip is not included in this analysis, although the data must be corrected for regional dip. Table 6-1 illustrates the dips on back and front limbs of a duplex, based on the number of thrusts that have been stacked (Suppe 1983). A fold formed by three thrust with a fundamental cutoff angle of 20 deg and containing three back limb dips is shown in Table 6-1. The fundamental cutoff angle Θ is shown in the central column, next to the folds forward and back dip domains. The numbers at the top of the table indicate the order of the dip domain and the number of thrusts. For a 20 deg cutoff angle, the back limb beds dip at roughly 20 deg, 38 deg, and 53 deg, indicating that the duplex contains three thrusts. The structure has the potential of repeating a reservoir unit three times, resulting in a *Third Order Duplex*. This procedure is referred to as **Dip Spectral Analysis.**

Dip Spectral Analysis

Forward dips							Fundamental cutoff angle° θ	Back dips						
VII	VI	V	IV	III	II	I		I	II	III	IV	V	VI	VII
61.6°	52.5°	43.0°	34.0°	25.2°	16.6°	8.2°	8°	8°	15.9°	23.4°	30.6°	37.3°	43.5°	49.3°
70.2°	59.2°	48.6°	38.3°	28.3°	18.6°	9.2°	9°	9°	17.8°	26.2°	34.0°	41.3°	47.9°	53.9°
80.6°	67.6°	55.2°	43.3°	31.9°	20.9°	10.3°	10°	10°	19.7°	28.9°	37.4°	45.1°	52.0°	58.2°
93.1°	77.3°	62.6°	48.8°	35.7°	23.3°	11.4°	11°	11°	21.6°	31.5°	40.6°	48.7°	55.9°	62.2°
109°	88.8°	71.0°	54.8°	39.8°	25.8°	12.6°	12°	12°	23.5°	34.1°	43.7°	52.1°	59.5°	65.9°
128°	102°	80.5°	61.5°	44.3°	28.5°	13.8°	13°	13°	25.4°	36.7°	46.7°	55.4°	62.9°	69.4°
160°	119°	91.3°	68.6°	48.9°	31.2°	15.0°	14°	14°	27.2°	39.1°	49.5°	58.4°	66.1°	72.5°
—	146°	104°	76.3°	53.6°	33.9°	16.2°	15°	15°	29.1°	41.5°	52.3°	61.4°	69.0°	75.5°
	—	124°	85.9°	59.0°	36.8°	17.4°	16°	16°	30.9°	43.9°	54.9°	64.1°	—	
		—	99.2°	65.6°	40.2°	18.8°	17°	17°	32.7°	46.2°	57.5°	—		
		—	123°	73.1°	43.7°	20.2°	18°	18°	34.4°	48.4°	59.9°	—		
			—	82.2°	47.4°	21.6°	19°	19°	36.2°	50.6°	—			
			—	97.6°	52.0°	23.2°	20°	20°	37.9°	52.7°	—			
				—	57.0°	24.8°	21°	21°	39.6°	—				
				—	63.6°	26.6°	22°	22°	41.3°	—				
				—	72.0°	28.4°	23°	23°	42.9°	—				
					—	30.4°	24°	24°	—					

Table 6-1 Dip Spectral Analysis. Each dip domain or distinct bed dip results in the possibility of an additional repeated section. Lowest bed dip defines the fundamental cutoff angle (central column). Right side of table applies to back limb dips, whereas, left side of table applies to front limb dips. Three distinctly different back limb bed dips define a Third Order Duplex. (From Suppe, Geometry and Kinematics of Fault-Bend Folding, 1983. Reprinted by permission of the American Journal of Science.)

The data for this analysis must be corrected for regional dip. In order to correct the data set for regional dip, subtract the regional dip from the observed dips. If the regional dip in the area being studied is 5 deg and the other dips are 25 deg, 43 deg, and 58 deg, then 5 deg is subtracted from the each dip. This results in regionally corrected dips of 20 deg, 38 deg, and 53 degrees. The lowest dip (20 deg), which is assumed to be the same as the fundamental cutoff angle, is found in Table 6-1. By consulting the right side of the table, which applies to back limb dips, we find that 20 deg, 38 deg, and 53 deg dipping beds constitute a match to a *Third Order Duplex*. This analysis can then be repeated on data from the frontal limb of the structure, which contains steeper dipping beds. The left side of Table 6-1 applies to the front, or more steeply, dipping limb. Dip Spectral Analysis can be used either as a QLT or applied to detailed analysis.

Consider the example shown in Figure 6-20a. The data consist of surface dips and a well which penetrates a flat thrust fault. Five different dips are shown ranging from 0 deg to 33 degrees. With the data provided, can we conduct Dip Spectral Analysis on the subsurface structure and determine the potential for repeated strata?

Before the Dip Spectral Analysis can be attempted, we may need to know something about the structural style of the area. The predominant structures for this area are fault bend folds and fault bend fold duplexes generated by ramp and flat thrust faults. With this general knowledge we can now analyze the available dip data. An analysis of the surface dip data suggests that a fold exists with the crest near the area of zero bed dip. Typically, we can assume that the lower angles of dip are related to the back limb of the fold, while the higher angles are related to the front limb. Therefore, in this example, the back limb is to the right of the zero dip area and the front limb to the left. The back limb has two different bed dips of 15 deg and 29 degrees. Likewise, the front limb has two dips of 16 deg and 33 degrees.

Dip Spectral Analysis can now be used to evaluate the potential subsurface structure shown in Figure 6-20a. First, make the assumption that the two front and back limb dips are indicating a duplex structure. Next, assume that the 15 deg dip represents the initial back limb dip, which would also represent the fundamental step-up angle of the fault. Table 6-1 is used to analyze the dips. For a fundamental cutoff angle of 15 deg, the back limb dips for a second order duplex are 15 deg and 29.1 degrees. The front limb dips for a Second Order Duplex are 16.2 deg and 33.9 degrees. These dips compare very well with the surface dips observed in Fig. 6-20a of 15 deg and 29 deg for the back limb, and 16 deg and 33 deg for the front limb.

Therefore, we can conclude from Dip Spectral Analysis that the structure is probably a Second Order Duplex.

Figure 6-20b illustrates a partially completed cross section using surface dip data. On this section additional well, dip and fault data are provided and the axial surfaces separating the various domains of constant dip are constructed. Additional work results in the completed interpretation of the second order fault bend-fold duplex shown in Figure 6-20c. This final example illustrates how Dip Spectral Analysis can be used as a QLT to evaluate a structural interpretation and the potential for penetrating repeated reservoir rock.

GROWTH

In simple terms, growth can be defined as the lateral change in sedimentary thickness across an area. Growth is an important, yet not well understood, subject. First, if a structure was **not growing** prior to or during the hydrocarbon generation and migration stage, then **no structure** existed at the time of migration to trap the hydrocarbons. Thus, it is very important to determine when a potential trapping fold began to grow. Second, data from the United States Gulf Coast, (Rainwater 1963; Fisher and McGowen 1967), Brunei, and elsewhere suggest that, within extensional terrains, large hydrocarbon accumulations are commonly discovered within depositional systems that exhibit high growth. We have some evidence that these relationships may also apply to some compressional terrains, although many well constrained examples supporting this contention are not presently available. However, the tectonostratigraphic process differs from extensional to compressional regimes.

Growth rates on extensional rollover structures are controlled by deltaic slumping or crustal extension, while growth rates on compressional structures are controlled by plate tectonics. Nevertheless, growth correlates to the hydrocarbon migration stage, to periods of rapid sedimentation, tectonic motion, and faulting. If hydrocarbons migrate along faults, then hydrocarbons are likely to migrate along growing, active faults into growing anticlines. Notable examples of production from compressional growth anticlines are present in Venezuela, Southern Oklahoma, the Newport-Englewood trend, and the Santa Barbara Channel in California.

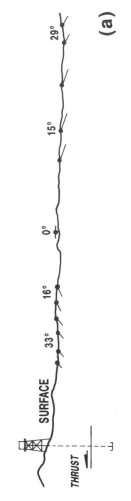

Figure 6-20a Cross section showing thrust in well and surface dipmeter data. (Modified from Suppe 1985. Published by permission of Prentice-Hall, Inc.)

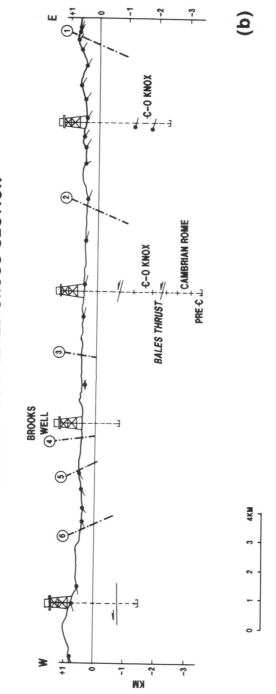

Figure 6-20b Partially interpreted section incorporating additional well log data. (Modified from Suppe 1985. Published by permission of Prentice-Hall, Inc.)

COMPLETED CROSS SECTION

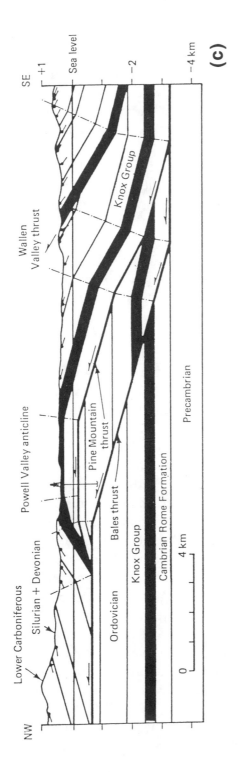

Figure 6-20c Completed interpretation. (From Suppe 1985. Published by permission of Prentice-Hall, Inc.)

In this section, we outline a method for determining when folds began to grow. The detailed theory required to explain our quick look techniques related to growth is beyond the scope of this book, requires an extensive discussion of fold kinematics, and is presented in other works including Suppe et al, (1992), Tearpock and Bischke (1991), and Xiao and Suppe (1992) for extensional environments. Any one interested in the details of the methods can consult these references.

We attempt, here, to simplify the more detailed analysis. For example, in compressional areas, symmetric, fault bend folds form as hanging wall beds, move up thrust ramps onto flats, and deform into anticlines (Rich 1934; Suppe 1983, 1985; Boyer 1986; Mitra 1986, 1988; Tearpock and Bischke 1991). This is shown in Figure 6-21. Initially, or before deformation, the hanging wall beds existed on the same structural level as the footwall beds (Fig. 6-21a). If the hanging wall slips over the footwall, the hanging wall beds move over *bends* in the fault surface. The bends in the fault surface cause the hanging wall beds to fold along axial surfaces. In Figure 6-21b, axial surfaces are shown to form at the convex and concave bends in the fault surface (Fig. 6-4).

As the deformation progresses, the hanging wall beds move up the thrust ramp and onto the upper flat (Fig. 6-21b). Gravity causes the hanging wall beds to deform and bend downwards forming the front limb of the fold. This deformation occurs at the convex bend in the fault surface (Figs. 6-21a and 6-21b). Similarly, beds moving through the concave bend and up the ramp in the fault surface form the back limb of the fold (Fig. 6-21b).

Let's now consider syntectonic sedimentation. If *syntectonic or growth sediments* are deposited across the limbs and crest of a fold, then these sediments will be thicker across the synclinal areas and thinner across the crest of the fold, as is commonly observed (Fig. 6-21c). As the fold moves forward during *a second growth stage*, sediments deposited on the fold crest, near the front limb of the fold, move off the crest and onto the front limb. This is shown as the heavily stippled region in Figures 6-21c and 6-21d. In other words, the thin sediments deposited on the fold crest during the first growth stage move onto the frontal limb of the structure during the second growth stage. At this stage, the geometry of growth folding begins to acquire an asymmetry. Similarly, as the fold amplifies and moves forward, growth sediments deposited on the back limb during the first growth stage (Fig. 6-21c), move partway up the ramp (Fig. 6-21d). The reason why the growth sediments deposited in the first growth stage move partway up the ramp is that these sediments did not exist when the pre-growth sediments first moved up the ramp. Motion of the pre-growth and growth sediments up the ramp causes the back limb dip domain to widen and the growth sediments to acquire an asymmetry (compare Fig. 6-21c to Fig 6-21d). In Figure 6-21e, we show a third growth stage which further adds growth sediments across the structure.

INITIAL PRE-GROWTH STATE

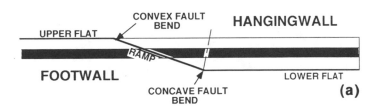

(a)

FIRST GROWTH STAGE

SLIP 1

(b)

DEPOSIT GROWTH SEDIMENTS

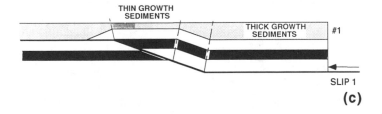

SLIP 1

(c)

Figure 6-21 Formation of a growth structure during folding. (a) Initial pre-growth ramp and flat geometry. (b) Hanging wall moves up ramp forming a fold. (c) Syntectonic sedimentation results in thin sediments deposited on fold crest and thicker sediments on fold limbs. (Modified from Suppe, Chou, and Hook (1992). Published by permission of John Suppe.)

This syntectonic sedimentation process forms growth sedimentary triangles and growth axial surfaces that do not bisect the growth sediment bed

152

dips, and imparts a characteristic asymmetry to the deformation (Fig. 6-21e). Thus, we can distinguish *pre-growth sediments, that have axial surfaces that bisect, from growth sediments that have axial surfaces that do not bisect the syntectonic sediment bed dips* (Fig. 6-21e and 6-21f). This is also shown on a seismic section from Southern California (Fig. 6-22).

SECOND GROWTH STAGE

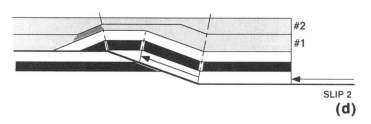

(d)

THIRD GROWTH STAGE

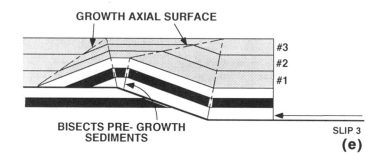

(e)

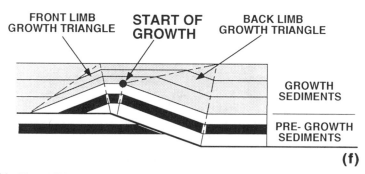

(f)

Figure 6-21 (d) Additional slip on fault combined with sedimentation results in an amplified growth anticline. Growth sediments deposited during first growth stage move up ramp, while thin sediments deposited on crest move onto front limb. Slip on thrust fault results in a characteristic asymmetry within the growth sedimentary section. (e) Additional syntectonic sedimentation results in formation of Growth Axial Surfaces that do not bisect bed dips within the growth section. (f) Process forms distinctive growth sedimentary triangle that can be used to date initiation of folding. (After Suppe, Chou, and Hook (1992). Published by permission of John Suppe.)

We saw in the earlier section on "Thin-Skinned Tectonics" that axial surfaces, which bisect fold limbs in pre-growth intervals, separate the individual dip domains. Indeed, it is an axial surface which forms or defines the boundary between two adjacent or contiguous dip domains. Figure 6-21b shows that axial surfaces only bisect the fold limbs within the pre-growth section or that part of the sedimentary section that formed prior to fold growth and amplification. Therefore, axial surfaces that do not bisect the bed dips must be in the growth sedimentary section. Thus, a parallel glider or a trained eye can rapidly locate the initiation of growth. In Figures 6-21f and 6-22, fold growth initiates where the growth axial surface terminates into a pre-growth axial surface.

Rapid Sedimentation Rates

In regions where growth sedimentation is rapid relative to fault slip rate, triangular growth patterns exist across anticlines (Figs. 6-21f and 6-22). The position where the growth axial surface terminates into a pre-growth surface defines the initiation of growth (Fig. 6-21f). In Figure 6-22, the axial surface below 1.05 sec bisects the fold limbs and defines the pre-growth sediments. Point G marks the onset of fold growth and defines the time at which the fold began to grow. Sediments deposited above Point G define the growth or syntectonic sedimentary section. Notice how the axial surface defined by *GAS* **(Growth Axial Surface)** does not bisect the deformed growth sediments. A protractor may be required to detect these differences. Therefore, if an axial surface is observed on a seismic section which obviously does not bisect the fold limbs (Fig. 6-22), then these sediments can be interpreted as growth sediments. Remember, the section must be roughly on a scale of one-to-one to apply this technique.

Empirical observations from extensional regions suggest that potential productivity is often related to the slope of the *GAS* (see Chapter 7). In the United States Gulf Coast, Brunei, and elsewhere, the shallower the dip of the *GAS,* the higher the growth and the better the potential productivity of the section. Conversely, a steeply dipping *GAS* is indicative of a lower growth and a poorer prospective section (Fig. 6-23). This conclusion is based on the observation that active fault surfaces provide an open conduit for migrating hydrocarbons (Leach 1993). Fault surfaces that are not active may seal rapidly.

154

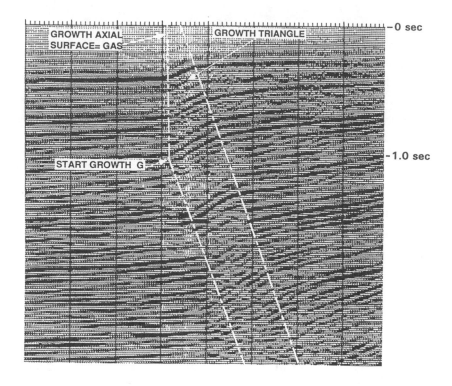

Figure 6-22 Growth triangle imaged at the front of a thrust structure. Growth initiates at Point G, where the axial surfaces at the front of the structure converge. Notice the axial surface within the growth section does not bisect the beds, while axial surfaces within the pre-growth section bisect the beds. (Seismic profile provided by Texaco USA and Nippon Western US Company Ltd. Published by permission of Chapman & Hall. Interpretation by John Suppe.)

Figure 6-23 shows how the slope of the *GAS* changes with respect to slip or slip rates. Observe that the ratio of fault slip (S) to sedimentary thickness (T), or the amount of forward fold motion, is larger in Figure 6-23a than in Figure 6-23b.

$$\frac{S_{fast}}{T_1} > \frac{S_{slow}}{T_2}$$

where:

T = sedimentary thickness or time
S = fault slip

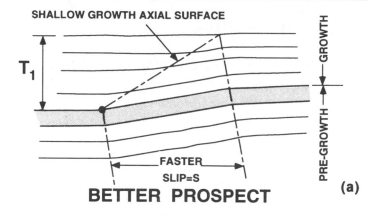

DETERMINING PROSPECTIVITY FROM GROWTH AXIAL SURFACES

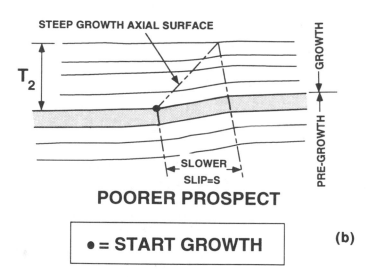

Figure 6-23 (a) Growth axial surface created by a fast growing structure, relative to the sedimentation rate, has a shallow dip. (b) Growth axial surface created by a slow growing structure, relative to the sedimentation rate, has a steeper dip.

If we assume that the sedimentation rates are similar in both cases, then the fold in Figure 6-23a is growing faster than the fold in Figure 6-23b. A high

156

pore pressure would cause the hanging wall to slip faster over the footwall. This faster motion creates a shallower dipping growth axial surface.

Slow Sedimentation Rates

In the previous section, we analyzed the case of high sedimentation rates relative to fold amplification or fault slip rate. These relationships have been recently discovered (Suppe and Medwedeff 1990; Suppe et al. 1992). Notice, in Figure 6-23a, that if the sedimentation rate were to slow substantially or the fault slip rate were to increase, then the *GAS* would *subparallel the top of the pre-growth sediments*. If the growth axial surface approaches the bedding, then this would create an onlap of the growth sediments (Fig. 6-24). This means that the fold is growing very rapidly relative to the sedimentation rate. The net effect is a fold that grows above the surrounding sedimentary basin, causing strata to onlap the front of the fold (Fig. 6-24). This onlapping relationship can also be used to date the age of folds and the onset of deformation. The fold began to form at the time of the oldest sediments that overlap the structure.

DETERMINING GROWTH INTERVAL FOR VERY LOW SEDIMENTATION RATES

START GROWTH

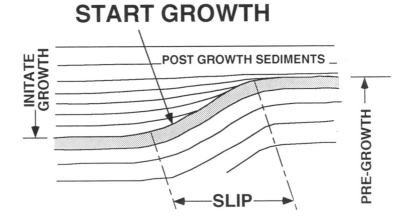

Figure 6-24 Very slow rates of sedimentation result in onlap of the units onto the rapidly rising structure. Growth is defined by the initiation of onlap.

QUICKLY EVALUATING "BALANCED" CROSS SECTIONS

You may have encountered balanced sections during prospect reviews. When reviewing prospects presented with balanced cross sections, how do you quickly evaluate whether the sections really are balanced? There are a few QLTs that can be applied.

In the introduction to this chapter, we presented two QLTs that can be used for checking balanced sections. The first is to check formation lengths, and the second to check formation thicknesses. These two simple techniques help evaluate so called "balanced" cross sections. Whenever a cross section is evaluated, first examine formation thicknesses. If the units or formations rapidly change thickness, then inquire as to what is causing the change in layer thickness. Be skeptical of answers that resemble, "This is what is observed in nature" or "This is what everyone does". Typically, only incompetent thick shale or evaporite horizons exhibit large thickness changes. In conclusion, the bed length and thickness rules are powerful QLTs that will aid prospect evaluation.

In the section, "Do The Axial Surfaces Bisect Fold Limbs?", we presented a simple relationship of fold geometry that makes the quick evaluation of cross sections possible. The limbs of symmetric or fault bend folds roughly bisect. If the limbs of asymmetric folds do not bisect within about 10 deg, or about twice the typical departure angle (Fig. 6-8), then you should ask for good evidence to justify why the structure is not obeying this rule. Numerous examples of productive structures that roughly obey the bisection rule are presented in Rich (1934), Suppe (1983, 1985), Boyer (1986), Mitra (1986, 1988), Suppe and Medwedeff (1990), and Tearpock and Bischke (1991). Unfortunately, the untrained eye may not properly detect subtle angle differences. However, a protractor can be used to quickly check for angle differences.

Balanced Section Pitfalls

In this section, we present several examples of potential pitfalls that may be encountered on "balanced" cross sections. Remember that balancing is no different than any other technique. It must be undertaken with the use of all available data, the correct methods, and conform to three-dimensional spatial relationships.

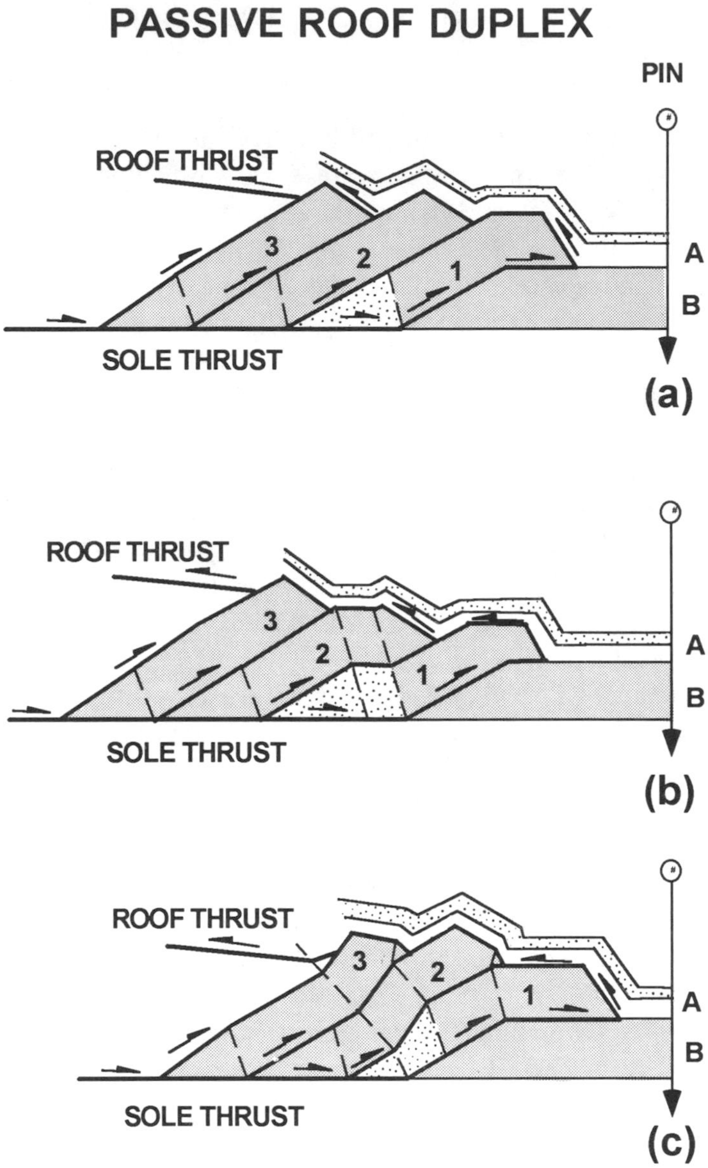

PASSIVE ROOF DUPLEX

Figure 6-25 (a) Passive roof duplex interpretation of fold and thrust belt. Large anticlinal structures form a simple piggy-back structural style. (Reprinted with permission from Journal of Structural Geólogy, v. 8, C.J. Banks and J. Warburton, *"Passive-roof" duplex geometry in the frontal structures of the Kirthar and Sulaiman Mountain Belts,* 1986, Elsevier Science Ltd, The Boulevard, Langford Lane, Kidlington 0X5 1GB, UK) (b) Reinterpretation showing how less motion along sole thrust creates flat dips within the duplex. (c) In another reinterpretation of 6-25a, additional motion along sole thrust creates the complex structural style commonly observed on seismic sections across fold and thrust belts.

Passive Roof Duplex

The interpretation shown in Figure 6-25a, employs the concept of a passive roof duplex (after Banks and Warburton 1986). Such interpretations are becoming popular on cross sections. According to this interpretation, thrust faults exist along the base and top of the thrust complex, causing the shallow units to remain unfaulted. Although we certainly believe that passive roof duplexes exist (Jones 1982; Boyer and Elliot 1982; Banks and Warburton 1986), the simple piggy-back stacking of the thrusts shown in Figure 6-25a involves a coincidence that is not likely to be reproduced in nature.

Notice in Block Nos. 1, 2, and 3 that the axial surfaces, which emanate from the bends in the thrust, outline a **wedge shaped** region between the dipping thrusts (Fig. 6-25a). This wedge shaped region is stippled in Block No. 1. A region of identical size and geometry occurs within every structural block. These blocks, which are totally surrounded by faults, are called **horse blocks**.

Notice that in Figure 6-25b, if less material moves up the frontal ramp, then the axial surface emanating from the bend in the fault surface would define a **trapezoidal shaped** horse block. This would create **flat dips** within the duplex (see Horse No. 1 and No. 2 in Fig. 6-25b). Furthermore, if the stippled portion of the horse block in Figure 6-25a moves partway up the ramp, then the straight faults shown in Figure 6-25a would be deformed (Fig. 6-25c), causing the structurally higher thrust faults to dip at higher angles than deeper faults. Typically, the dip of the higher thrust is slightly less than twice the dip of the structurally lower thrust fault (Suppe 1983). This topic was described in the section on "Duplex Structures".

The interpretation shown in Figure 6-25a is generated from a simple structural concept. Geoscientists may generate from minor changes in the surface deformation, large subsurface structures that may be unrealistic.

Fish Hook Faults

Any correct interpretation can be restored or retrodeformed to its initial or undeformed state (Woodward et al. 1985; Marshack and Mitra 1988). You may recognize this concept as palinspastic restoration. The ability to restore a section provides a check on the interpretation. Some people who balance cross sections restore their interpretations to an initial state. When evaluating these interpretations, the restoration is often more informative than the "balanced" cross section. For example, Figure 6-26a is a cross section of a duplex structure modified from the literature. The interpretation appears reasonable and several potential plays are present. However, the restored section generates several **"fish hook"** faults (Fig. 6-26b). A "fish hook" fault is the compressional analog of the extensional "saucer" fault described in Chapter 4. Why did the thrust not take the path of least resistance and rupture upward like the other thrusts (Fig. 6-26c)? After considering the fish hook faults resulting from the restoration, it seems that the interpretation shown in Figure 6-26a is not physically reasonable and should be modified.

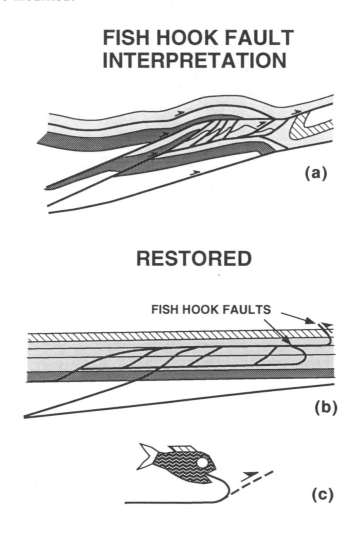

Figure 6-26 (a) Reasonable looking interpretation of a duplex structure. (b) Restoration of the faults to their initial state causes the frontal faults to curve back forming "fish hook faults". (c) Thrust faults typically rupture upward at a fundamental cutoff angle.

Multi-Play Pitfall

Figure 6-27a shows an interpretation that is sometimes encountered in thrust belts. From well log correlations, a repeated section and a thrust fault are recognized in each of the three wells. The thrusts are assumed to ramp off a main décollement as three different thrusts. However, a coincidence occurs in that the thrusts do not intersect the adjoining wells (Fig. 6-27a), nor do they breach the surface.

MULTI-PLAY INTERPRETATION

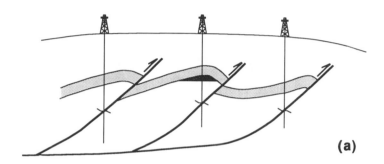

(a)

ALTERNATE INTERPRETATION

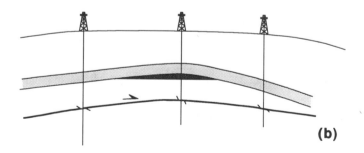

(b)

Figure 6-27 (a) Multi-play pitfall. Well log correlations from three wells indicating three repeated sections, with only one thrust fault in each well. On this evidence, three prospects are created along three thrusts faults. Notice that the thrust faults do not intersect the adjoining wells. Is this just a strange coincidence? (b) Interpretation of the same data using a single thin-skinned thrust that is slightly folded. If this interpretation is correct, then there is only one discovery possible in this area.

An alternative interpretation involves the concept of **folded thrusts** between the wells (Jones 1971; see section on "Duplex Structures"). In other words, a single thrust that is slightly folded exists between the wells and thus there are *not* two additional structural plays waiting to be drilled (Fig. 6-27b).

FAULT PROPAGATION FOLD DIP ANALYSIS

Structural interpretations should conform to geologic and geometric principles in both cross-sectional and plan views. For example, when reviewing a structure map of a fault bend or fault propagation fold, it is possible to examine the accuracy of the map using forward modeling techniques.

The best way to illustrate forward modeling techniques is with an example. Consider the structure map in Figure 6-28 which shows a recent, undeveloped, discovery. The structure is interpreted as a fault propagation fold formed by a 35 deg reverse fault. Since a fault propagation fold is a fault related fold, there must be a geometric relationship between fold shape and fault shape. Such a relationship is shown in Figure 6-29. In this generic example, the fundamental fault cutoff angle is $\Theta = 25$ degrees. Applying fault propagation fold geometric principles (Suppe 1985) and using the fault cutoff angle, we can determine four corresponding fold angles γ_p, γ^*_p, β, and the back limb axial angle (see Figure 6-29b). The structure map of this generic fold should honor these fold angles.

The two primary angles to check first are the back and front limbs, which correspond to Θ (the fundamental cutoff angle), and $180° - \beta$, respectively. Let's review the structure map in Figure 6-28 by applying QLTs to the front and back limb dips. This can be done by first laying out the illustration aid cross section X-X´ through the structure, parallel to transport direction, as shown in Figure 6-30. The black dots represent the depths for each of the contours on the structure map. The cross section is an exact representation of the structural interpretation. The reverse fault shown in the figure comes from the fault surface map. From the cross section, we can see that the back limb dip is 35 deg, which conforms to the dip of the fault, which is 35 degrees.

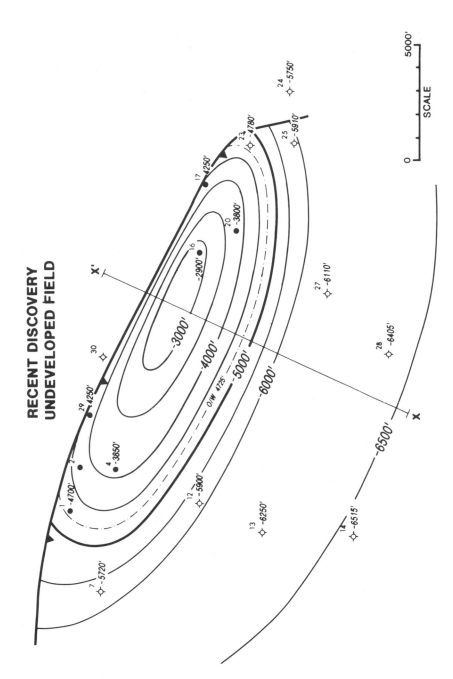

Figure 6-28 Structure map constructed from well log correlations across an asymmetric anticline that contains steep front limb bed dips. Cross section X-X′ is used to check structural balance.

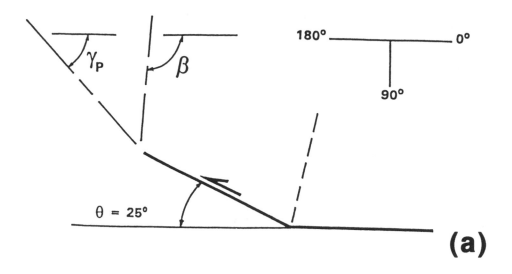

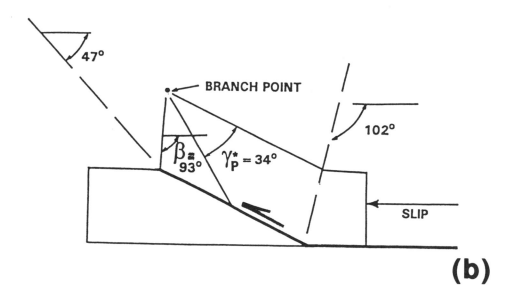

Figure 6-29 Construction of a generic fault propagation fold through four stages of development. (a) Blind thrust fault ramps off a 25 deg cutoff angle. (b) Construction of branch point and γ_p, γ^*_p and β angles.

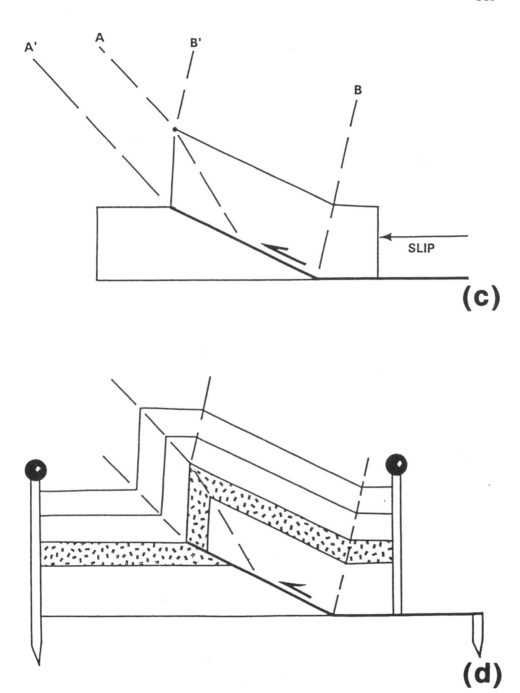

Figure 6-29 (c) Construction of framework. (d) Completed structure.

166

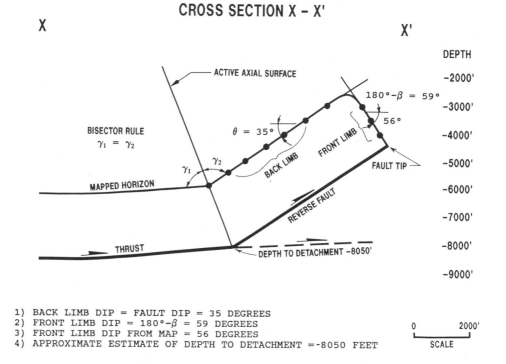

Figure 6-30 Illustration aid cross section along profile X-X´ of Figure 6-28.

We graphically measure the front limb dip from the map or cross section and compare this dip to that determined from fault propagation fold forward modeling. The front limb dip for this fault propagation fold is defined as:

Front Limb Dip = 180° - β
where $\beta = \Theta + 2(\gamma^*_p)$

Using the fundamental cutoff angle of 35 deg and the nomogram in Figure 6-31, we can determine the axial angle γ^*_p. Therefore, the front limb dip for this fold should approximately be:

Front Limb Dip $= 180° - [35° + 2(43°)]$
$$= 180° - 121°$$
$$= 59°$$

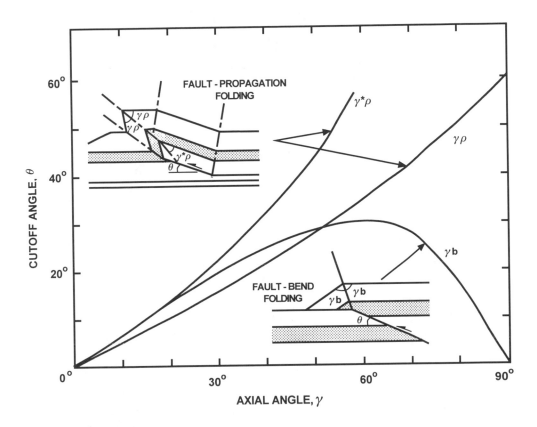

Figure 6-31 Fault bend and fault propagation fold nomogram. (Modified from Suppe 1985. Published by permission of Prentice-Hall, Inc.)

From the structure map or cross section, the front limb dip is measured as 56 deg, which matches favorably with the 59 deg calculated using forward modeling.

Based on this general analysis of fault propagation fold dips, the structural interpretation of the newly discovered field shown in Figure 6-28 appears reasonable. Such an analysis can also be conducted on mature fields to aid in the structural evaluation. A similar dip analysis can also be conducted on fault bend folds.

CHAPTER SEVEN

EXTENSIONAL STRUCTURAL GEOLOGY QUICK LOOK TECHNIQUES (QLTs)

INTRODUCTION

Extensional tectonics is fundamentally different from compressional tectonics, and one must change mind sets when changing tectonic regimes. Both regimes contain growth sediments (extensional more so than compressional); thus some of the basic principles carry over to normally faulted structures (i.e., growth is important). The two different structural styles, which image as fundamentally different fold shapes on seismic profiles, suggest a **different deformational process.** Thus, extensional tectonics is not the inverse of compressional tectonics. Similarly, compressional folding is not the reverse of extensional folding.

This chapter deals primarily with a growth faulted extensional structural style. This structural style results in the formation of rollover anticlines in the hanging wall block of large normal growth faults.

LISTRIC FAULT SHAPE DETERMINES ROLLOVER SHAPE

In Chapter 6, we saw that there is a relationship between the shape of the fault that formed the structure and the associated fold geometry. This basic principle also applies to rollover structures formed by growth normal faults. Using this principle, a number of Quick Look Techniques are available to evaluate prospects.

First, we present a **simple** model of rollover formation in order to show how this process works. Hamblin (1965) recognized that rollovers form as hanging wall blocks move along listric normal faults. Therefore, if the hanging wall block moves over the footwall block, a void or hole is opened up between the two blocks (Fig. 7-1). This hole is instantaneously closed by gravitational collapse of the hanging wall block upon the footwall (Gibbs 1983; Tearpock and Bischke 1991). In the example shown in Figure 7-1, we assume that the gravitational collapse occurs along *inclined shear surfaces*, as has been demonstrated by rock mechanics (Jaeger 1962). Xiao and Suppe (1992) present empirical evidence from a number of well constrained rollovers in support of these conclusions.

From empirical data and observations, we conclude that for a given amount of slip, a fault that is **more listric or flat** (Fig. 7-2), will displace the hanging wall a **smaller vertical distance** than a fault that has a steeper dip (compare Fig 7-1 to Fig 7-2). *In other words, the amount that the crestal beds are vertically displaced or dropped is small, producing a rollover high in map view.* If, on the other hand, the normal fault is **less listric** and dips at a steeper angle, then the hanging wall block will drop a **larger vertical distance** for the same amount of slip (Fig. 7-1). In this case, the amount that the crestal beds are vertically displaced is larger, creating a rollover low or saddle in map view.

Large rollover highs are often located over those portions of listric normal faults that are *bowed out or flat* (Fig. 7-2). The flanking saddles that laterally close large rollover highs (Fig. 7-5) are located over those portions of normal faults that are less listric. Saddles form over fault surfaces with high gradient, referred to as *chutes* (Fig. 7-2). Therefore, large rollovers are closed along strike by listric faults that dip at a steeper angle, relative to the flats located beneath large rollover highs. You may wish to consult Tearpock and Bischke (1991) for documented examples.

RAPIDLY LOCATING MAJOR ROLLOVER HIGHS

The following discussion is directed at large growth fault systems, like those associated with large deltas. Similar, although less dramatic, features are also observed in regions of crustal extension.

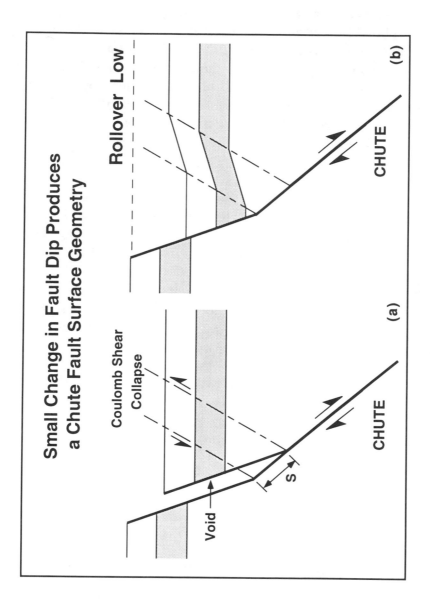

Figure 7-1 (a) Hanging wall moves over footwall and then (b) collapses to fill void. Collapse occurs along shear surface pinned to bend in fault surface. If fault is very listric or "flat", then a rollover high is generated.

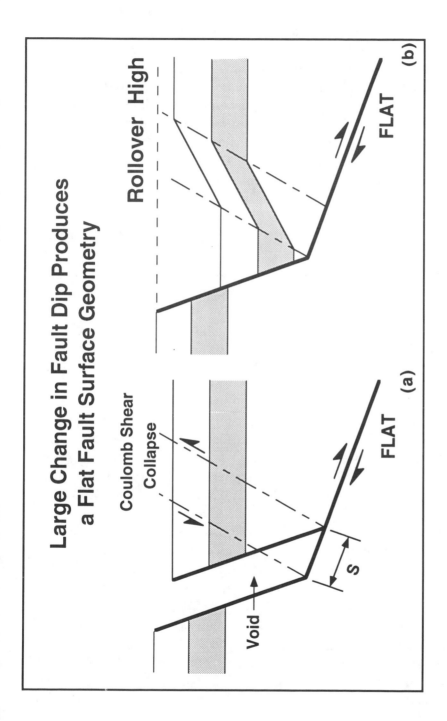

Figure 7-2 (a) If the fault surface dips at a steeper angle or along a "chute" geometry, then (b) the rollover will have a lower amplitude forming a flanking syncline. Thus the geometric configurations shown in Figures 7-1 and 7-2 can be used to locate rollover highs.

In many areas, large rollovers are associated with one major expansion fault, and often with one or more synthetic faults that exhibit less vertical separation than the major expansion fault. In Figure 7-3, the major expansion fault is imaged between sp A and D (surface to 3.5 sec.). In many cases, the shape of the fault surface, as discussed in the previous section, can be used to steer you *directly to a rollover prospect*, and allow you to quickly check its structural viability. These techniques are fundamental to proper prospect evaluation and generation in extensional growth fault environments. If however, the prospect is influenced by several large synthetic faults that exhibit about the same expansion, then these large synthetic faults may behave similar to a major fault and the analysis will require more detailed study.

First, let's take a quick look at a few dip oriented seismic lines. As we have mentioned, many rollovers are associated with faults that are bowed or curved. Thus, the observations discussed in the last section on Listric Fault Shape suggest a method for rapidly locating large rollover highs, and for checking prospect maps for structural viability. Start by organizing all of the dip lines from a large prospect and arrange them in consecutive order along strike. Managers may simply choose to examine the montage lines for this analysis. Next, locate the large expansion fault that formed the rollover (Tearpock and Bischke, 1991; Xiao and Suppe, 1992). The fault that formed the rollover exhibits the largest expansion relative to the other synthetic faults, and is often well imaged on each profile. On the two seismic lines shown in Figures 7-3 and 7-4 the fault that formed the rollover stands out as a booming reflector between sp A and D (between 2.0 and 3.5 sec.) and sp A and D (between 2.0 and 3.4 sec.), respectively. Using these dip lines compare the relative fault dips of the major fault beneath the crest of the rollover.

In Figure 7-3 the major fault is flatter (dips at a more gentle angle) beneath the crest of the rollover (sp B and C), than it is in Figure 7-4 (sp B and C). This means that the rollover high is located near Figure 7-3. In Figure 7-4 the major fault has a **steeper fault dip** and is below a saddle or syncline that laterally closes the rollover high (Fig. 7-5). So in general, seismic profiles located across flats or bows in a fault surface correspond to the position of **rollover highs** (Figs. 7-2 and 7-3). These profiles exhibit *flatter fault dips* beneath rollover crests than saddle profiles (Figs. 7-1 and 7-4). Rollover lows (Fig. 7-4) correspond to relatively *steeper fault dips*. It is the rollover low or saddle that laterally closes the structure (see Fig. 7-5).

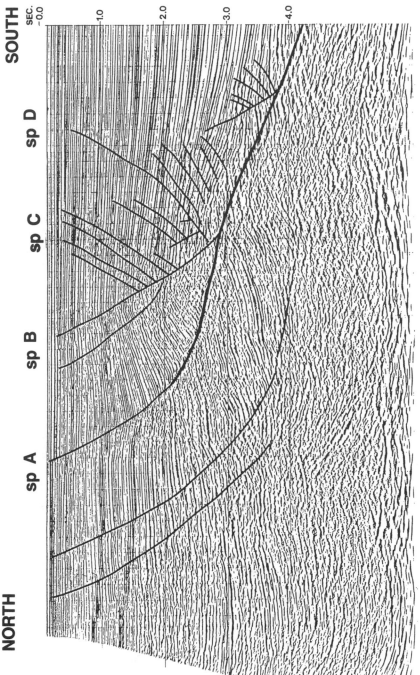

Figure 7-3 Bow fault geometry located across a large rollover high. Compare "flatter" fault dips beneath roller crest to the chute geometry imaged in Figure 7-4. Flatter fault dips indicate this line crosses a major rollover high (Figs. 7-1 and 7-5). (Seismic Published by permission of JEBCO Seismic, Inc., Houston, Texas. Interpretation by Richard Bischke.)

174

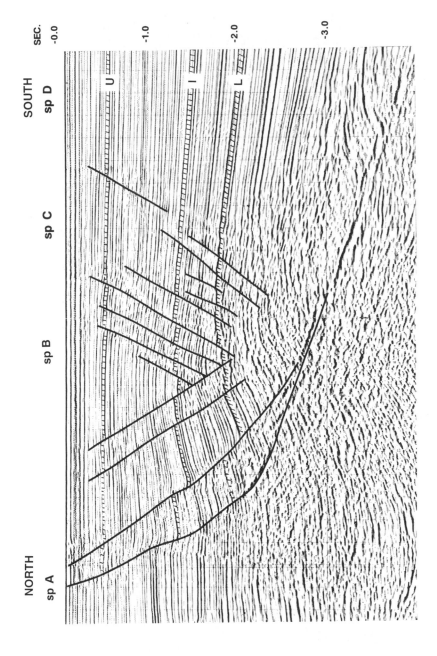

Figure 7-4 Chute fault geometry. Notice how the fault does not "flatten" beneath rollover crest. This fault geometry indicates a major rollover low, or syncline, that laterally closes the rollover (Figs. 7-2). (Seismic published by permission JEBCO Seismic, Inc., Houston, Texas. Interpretation by Richard Bischke.)

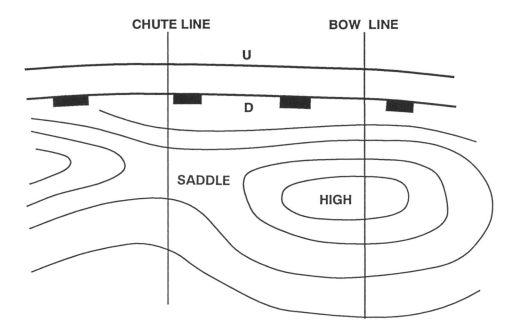

Figure 7-5 Fault "flat" geometry crosses rollover high, while chute geometry crosses flanking syncline (see Figs. 7-1 to 7-4).

The relationship of fault shape to rollover shape is very important to prospect evaluation. If these flatter and more steeply dipping fault geometries do not correspond respectively to the highs and lows on a structure map, then critically examine reflection picks. There is likely to be a correlation bust or structural interpretation problem.

Let's take a quick look at a *strike oriented seismic line*. Tearpock and Bischke (1991) discuss the reasons why some geophysicists mistrust or ignore strike lines. Although strike lines may exhibit side-swipe and need a more critical evaluation, strike lines can be used effectively to interpret structure.

We believe that, often, more can be learned from strike lines than from dip lines. This is based on our experience, as well as that of our associates, and includes the discovery of several billion barrels of oil, worldwide. *One of the major pitfalls of structural interpretation is the improper use of strike lines.*

We contend, that if a structure is *not present* on well processed strike lines, then it will not be present on dip lines either. For example, examine the well processed strike line shown in Figure 7-6. The beds at shallow depths (sp B to E at 1 to 2 sec) are conforming to the shape of the fault located at 2.3 to 2.8 sec. At the 2.3 to 2.8 sec level, notice how the fault is *flat* beneath the crest of the rollover (sp C to D) and how the fault turns down along the flanks of the rollover (sp B to C and D to E). Notice that the beds above the fault mimic the shape of the fault. *Again, we see that fault shape determines fold shape, on strike lines, as well as on dip lines.* Using the strike line, we can now be fairly confident that a major rollover has been located. This conclusion should be confirmed on dip lines and by mapping several horizons.

In large growth fault systems, the rollover may exist over large vertical distances (Figure 7-3 and 7-4). The single strike line shot between sp A and D (Figs. 7-3 and 7-4) images the rollover. Now that the rollover has been confirmed, the next step is to *quickly determine rollover size.* Notice that a single strike line images the lateral extent of the rollover more accurately than several dip lines crossing the same structure (Fig. 7-6). This is because strike lines better image the *lateral dimension of a rollover*, which is normally greater than or equal to the dip line dimensions (Figs. 7-3 and 7-4). In other words, if a single dip line exists over a rollover high, then we know that the rollover exists. We also know that it is probably as wide, or wider, than the dip line dimension. But how wide is it? However, if a single strike line crosses a rollover high, we immediately know whether the structure is small, moderate in size, or a giant.

How can we better determine prospect shape and size? We have found that large rollovers often terminate rapidly in a lateral or strike direction. If this is not recognized, then the size of a prospect generated on the rollover may be overestimated. An example of a rapid lateral rollover termination is shown in Figure 7-6 between sp B and C. On this strike line, the beds above the fault are steeply dipping to the west between sp B and C. This means that dip lines shot across the flanks of this large rollover will record out-of-the-plane energy or side-swipe. The strike line, and not the dip line is better imaging the flank of the structure. On the flanks of a structure, regional strike lines often represent local dip lines with respect to the structure. If a strike line suggests steeply dipping flanking beds, then the structure map should be checked to see if the strike lines were correctly used to map the steeply dipping flank beds.

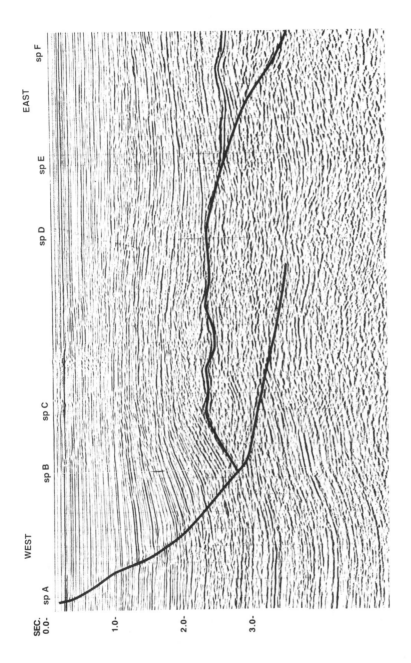

Figure 7-6 Strike line crossing a major rollover structure. The fault surface which formed the producing rollover (above 2.3 sec) is "flat" beneath the rollover crest (sp C to D, at 2.3 to 2.4 sec) and dips beneath the rollover saddles or lows located at sp B to C (2.85 to 2.3 sec) and at sp D to E (2.4 to 2.7 sec). (Seismic published by permission of JEBCO Seismic Inc., Houston, Texas. Interpretation by Richard Bischke.)

The use of strike lines has major implications when exploring deltaic systems in frontier areas (Tearpock and Bischke 1991). Millions of dollars are spent shooting dip lines across areas that do not contain rollovers. A few strike lines shot across the same area can confirm whether any structures exist. This can be done at a fraction of the cost of shooting a large number of dip lines. Strike line spacing can be determined by the minimum required prospect size.

Similar logic can be applied to an area that is known to contain rollovers. If rollovers exist in an area, then their presence will be imaged on the strike lines. Consequently, when considering an area for exploration, money can be saved by purchasing or shooting exploratory strike lines. Several dip lines can also be purchased for tie lines. If rollover structures are confirmed, then additional dip lines that cross the structural highs can be purchased or shot for detailed prospect mapping.

ANTILISTRIC FAULT BENDS: A SAND INDICATOR

Two types of normal faults are common on the crests of rollovers: (1) synthetic faults, which typically dip toward the basin and (2) antithetic faults, which typically dip landward or toward the master fault. In the previous section, we showed that synthetic faults tend to control the shape and size of large rollover structures; therefore, synthetic faults normally exhibit larger expansions than antithetic faults. Synthetic and antithetic faults can be either **listric**, exhibiting a shape that is *concave upward*, or **antilistric**, with a *convex upward* shape.

Antilistric faults are faults that steepen, their dip with depth and (Figs. 7-7 and 7-8) have been described by Ocamb (1961), Roux (1978), Bischke and Suppe (1990), Tearpock and Bischke (1991), and Bischke and Tearpock (1993). This particular fault shape is so important that it is difficult to understand why more emphasis has not been given to its recognition and understanding. Simply put, antilistric fault bends may be **sand indicators.** This concept is based on the well known fact that sand sections compact less than shale sections (Roux 1978; Xiao and Suppe 1989). Thus, the dip of a fault in a shale section is less steep than the dip of a fault in a sand section (Figs. 7-7 and 7-8).

Before outlining the procedure for using fault dip as a sand indicator, we list a number of limitations and problems associated with the method. In many areas such as the United States Gulf Coast, the North Sea, the Niger

Delta, and Brunei, the major, down-to-the-basin faults tend to flatten rapidly at depth. The reasons for this flatter or more listric fault shape are not entirely known, but the fault shape is **not** the sole result of sedimentary compaction. The antilistric sand indicator method applies to situations where *footwall compaction* is the dominant fault deformation process. Accordingly, the method applies best to the shallower portions of growth synthetic faults, and to growth antithetic faults that are not subject to other deformational processes such as diapirism (Bischke and Tearpock 1993). Because antithetic faults are not often observed to flatten at greater depths (as is observed along major synthetic faults), the technique does not appear to be depth limited with antithetic faults (Bischke and Suppe 1990). However, if the antithetics flatten with depth to angles that are lower than permitted by compaction, then these faults cannot be used to determine the presence of sand.

It is usually not obvious that antilistric normal faults are present on many seismic sections. This results from the fact that most seismic sections are generated on a scale that is *vertically exaggerated*. Vertically exaggerated (horizontally compressed) seismic sections have the effect of **straightening** out normal fault dips. We have previously demonstrated that distorting the true geometry of a structure represents one of the greatest pitfalls in structural interpretation (Tearpock and Bischke 1991). Depth corrected seismic data can show the geometric relationships present on seismic sections at a scale of one-to-one, provided the data are properly processed and migrated. In today's interactive computer environment, there are few reasons for not working at or near true scale. We have found that interval velocities can often be used to depth correct seismic data, although more accurate velocity functions are usually available. Even average velocities applied to the interval of concern will generate a clearer picture of the geometries present on the section. Cross sections constructed at near true scale are significantly better than sections constructed at an exaggerated scale.

The presence of sand is often a major risk in prospect evaluation. At times, the analysis of fault dips may provide important information about the presence of sand. Consider a growth normal fault that forms in a sand-shale section deposited near the sea floor (Fig. 7-7a). As sand sections have a lower initial porosity than shale sections (40% for sands vs. about 68% for shale sections) (Xiao and Suppe 1989), sands compact less than shales. For this reason, growth faults dip at lower angles over footwall shale sections, than over footwall sand sections (Fig. 7-7b).

COMPACTION ALONG A
GROWTH NORMAL FAULT

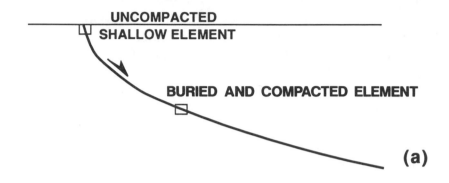

(a)

ANTILISTRIC FAULT GEOMETRY

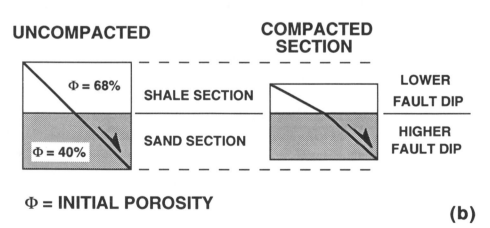

(b)

Figure 7-7 Compaction along a growth fault can be used to identify sand or shale sections. (a) Near surface element containing growth normal fault will experience compaction and fault rotation upon burial. The compaction applies to the footwall and not to the hanging wall block. (b) As shale sections have a higher initial porosity than sands, shales compact more than sands. Thus, growth normal faults in sand sections have higher dips than growth normal faults in shale sections. This results in an antilistric fault bend or sand indicator.

This principle has been used to *qualitatively* recognize the presence of sand for at least forty years (Gow 1962). Recently, a *quantitative* method was developed to estimate sand-shale ratios, or percent sand within **footwall** fault blocks (Bischke and Suppe 1990; Tearpock and Bischke 1991). In many growth faulted areas, hydrocarbons are typically found in the hanging wall or downthrown block. Therefore, it is also important to be able to estimate the sand percentage in the hanging wall block. This estimate can be made in many situations. Assume that a prospect is generated downthrown to a crestal antithetic fault. If the vertical separation across the crestal antithetic fault is small, then it should be easy to correlate from the footwall into the corresponding hanging wall beds. Furthermore, as the hanging wall block is downthrown, it probably contains thicker sands and has the possibility of a higher sand percentage than the equivalent upthrown section.

The antilistric or fault pattern is readily recognized on many seismic sections. Most seismic sections are, however, vertically exaggerated, which has the effect of straightening out the true geometry of the fault surface. If the seismic section is not depth corrected, then place a transparent straight edge or parallel glider along the shallower portions of the antithetic faults. This has the effect of isolating subtle changes in fault shape. The straight edge can show subtle changes in fault geometry that may not be obvious from an eyeball examination of the section. Several antilistric faults are seen on the conventional seismic line shown in Figure 7-3 between sp B and D, at about 1.0 to 1.3 sec. Locate the position where the antithetic faults turn downward and become antilistric, with increasing depth. This antilistric bend in the fault is an indicator that more sand is present in the footwall at this interval.

Remember, that changes in fault curvature are more readily identified on sections that are displayed at true scale. A depth corrected example, using interval velocities, is shown from a Brazos Ridge field, offshore Texas in Figure 7-8. In this example, a higher percentage of sand is present where the antithetic growth faults have a steeper dip, as shown by Fault No. 1 as it cuts Horizon A. The average sand-shale ratio from a nearby well is included on the figure. Notice how the faults become more listric in the section beneath Horizon B. This indicates that the section is getting more shaly with depth. When applying percent sand analysis, you should analyze several lines for consistency. Other examples of the use of antilistric faults as indicators of higher sand percent are presented in Gow (1962) and Roux (1978).

182

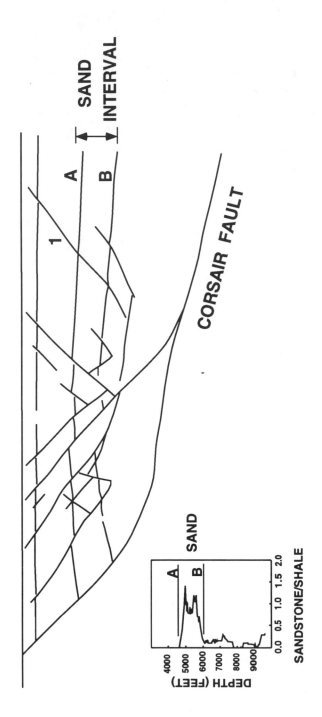

Figure 7-8 Digitized and depth corrected seismic section from along Corsair (Big Hum) trend, offshore Texas. Growth faults are antilistric into sand section. (From Bischke and Suppe (1990). Published by permission of Gulf Coast Association of Geological Societies.)

AXIAL SURFACES: LOCATING FAULT POSITION

In the next two sections, we present a number of qualitative QLTs that will allow you to rapidly determine whether fault interpretations are structurally correct. There seems to be a lack of concern regarding fault interpretations and fault mapping in our industry today. We find this surprising, since there is often a close relationship between faulting and the location and size of a prospect. Even in good data areas, or where the seismic reflections do not change character, seismic reflection analysis may or may not guide us to the correct fault interpretation. How do we interpret faults in areas where the data are **poor**, or where reflections loose resolution or change character?

There are many examples of very costly dry holes that have resulted from *incorrect fault interpretations*. In our opinion, if more attention were given to fault surface interpretations, the percentage of dry holes drilled on compressional and extensional structures could be significantly reduced. Hence, we stress the importance of analyzing and mapping faults.

In analyzing faults, we evaluate the overall fault interpretations, the fault shape and its position. These factors are important because:

1. fault shape and position control prospect size and location;
2. we need to know whether reservoir beds are in the hanging wall or the footwall;
3. the stratigraphy changes across fault surfaces;
4. they provide clues as to where to prospect for additional plays; and
5. correct fault interpretations keep us more focused, help us alleviate structural problems, and can reduce the number of dry holes drilled.

In Chapter 6, we discussed the importance of axial surfaces and how they can be used to *predict fault shape and position,* as well as correctly position wells on a prospect. Axial surfaces also exist within rollover structures, although they are often *hidden* in a maze of complex antithetic and synthetic faulting. Also, within the shallow portions of a growth section, where bed dips are low, the expression of axial surfaces is very **subtle**.

We now present several well imaged examples of axial surfaces on seismic sections and apply these features to prospect evaluation. What causes axial surfaces? An examination of Figures 7-1 and 7-2 shows that, as the hanging wall collapses onto the footwall, the beds rollover forming a trap.

Notice that the hanging wall collapse occurs along a shear surface that *emanates from the bend in the fault surface* (Tearpock and Bischke 1991; Xiao and Suppe 1992). The shear surface that causes the hanging wall collapse defines the axial surfaces. The axial surface that causes the active deformation is adfixed to the bend in the fault surface. In some ways, fault bends determine prospect position by controlling the angle at which the beds dip and the location of individual structural dip domains. This was discussed in Chapter 6. Consequently, if we can map the dip domains, this work can help us determine fault shape and possibly fault position.

How do we locate axial surfaces? Often, interpreters approach this subject by using the traditional interpretation method of sighting along deformed reflectors. This method is often inadequate and is probably responsible for not recognizing axial surfaces on seismic sections. A straight edge, rolling ruler, or parallel glider *should* be used during this analysis, because a *straight edge can project straighter than the eye*.

ALIGN RULER PARALLEL TO
COHERENT REFLECTION PACKAGE

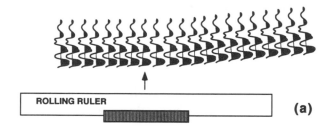

(a)

MOVE RULER OVER REFLECTIONS

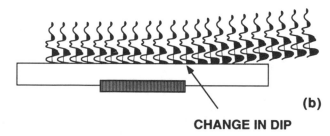

(b)

CHANGE IN DIP

Figure 7-9 Dip domain analysis. (a) Rolling ruler is aligned parallel to and then passed across coherent packages of reflectors defining the dip domains. (b) Change in bed dip is marked on seismic profile.

Figure 7-9 illustrates how to use a rolling ruler to identify and locate an axial surface. First, align a rolling ruler with a package or group of coherent reflections that do not significantly change dip. This is illustrated in Figure 7-9a. Next, pass the rolling ruler across the section, marking the positions where the reflections change dip, as shown in Figure 7-9b. These uniformly dipping packages of coherent reflectors are called dip domains (Suppe 1985; Tearpock and Bischke 1991). This process should then be repeated in the adjacent dip domain (Fig. 7-9c). *These changes in dip on the section, which separate different packages of reflectors having nearly constant dip, define the axial surfaces.* In other words, axial surfaces exist where uniform regions of nearly constant dip change to regions of different dip (Fig. 7-9d).

CONFIRM IN ADJACENT DIP DOMAIN

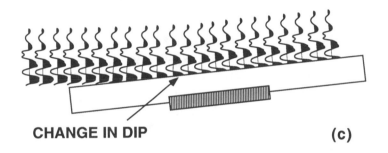

CHANGE IN DIP   **(c)**

DEFINING TWO DIP DOMAINS

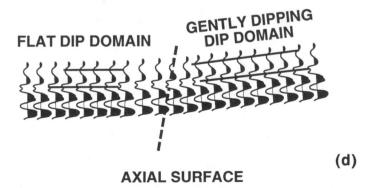

FLAT DIP DOMAIN   **GENTLY DIPPING DIP DOMAIN**

AXIAL SURFACE   **(d)**

Figure 7-9 (c) Process is confirmed in adjoining dip domain. (d) Procedure locates axial surface which can be used to define fault geometry and prospect size and location.

Axial surfaces may be very subtle at shallow depths, but are usually obvious at greater depths where the reflections are more steeply dipping or rolled over. Their location and trends are important, for *these surfaces define growth sections on seismic profiles* and can be used as an aid in determining fault shape and position as is discussed in the section on "Common Fold Patterns". After a little practice, this axial surface technique can be used to improve interpretations, thereby reducing the chances of drilling a dry hole.

RELATIONSHIP BETWEEN NORMAL FAULT SHAPE AND PROSPECT SHAPE AND POSITION

One of the most significant structural observations that can be made during the mapping and interpreting of rollover structures (particularly in poor data areas) is to notice the relationship between fault shape and rollover fold shape (Nunns 1991; Dula 1991; Xiao and Suppe 1992). Refer once again to Figures 7-1 and 7-2. If you know the shape of a fault, then you can predict the shape of the fold formed by the fault. Conversely, if you know fold shape you can predict the shape of the fault and its approximate position. If an interpretation under review is correct, then these two features, fold shape and fault shape, should be in rough agreement. If not, then the interpretation is likely to be in error.

Dry holes and economically unsuccessful wells have been drilled because this simple relationship was not understood or recognized. If you are questioning whether such a simple relationship exists, ask yourself, have you ever seen a listric growth fault that did not have an associated hanging wall rollover, or have frontal hanging wall bed dips that did not steepen with increasing depth? The answer to these questions is surely **"no"**. Therefore, some fundamental tectonic factor controls rollover development. This factor is **fault shape** (Tearpock and Bischke 1991; Xiao and Suppe 1992). A complete discussion of the mechanics of this relationship is beyond the scope of this book and is discussed in Xiao and Suppe (1992). However, you should be aware of the fact that *fault shape determines rollover shape, and thus, prospect size and location.*

Next, we address a common misconception that is sometimes responsible for dry holes. It seems that some interpreters have been taught that loading at the front of a rollover is solely responsible for the formation and geometry of a rollover, including the observed fanning of bed dips above the listric bends in the major fault. The proposed mechanism responsible for this deformation is ascribed to block rotation along a listric normal fault. However, this simple loading and rotation hypothesis ignores **volume conservation principles** (Goguel 1965). For example, if the hanging wall block simply rotates over a listric fault (Fig. 7-10a), then the rollover would not close or reverse its dip at its crest, and a void would develop along the lower portion of the fault surface (Fig. 7-10b). Gibbs (1983) showed by simple geometric grounds that block rotation causes *unbalanced* and *inadmissible* cross sections. A simple block rotation requires that the underlying fault be circular and that its basinal extension be a reverse fault. The block rotation explanation is too simplistic. If this incorrect concept is applied to a poor data zone or complex structural problem, it may lead to a prospect that results in a dry hole, or, at least, a confusing problematic interpretation that eventually will require reinterpretation.

ROTATED BLOCK HYPOTHESIS

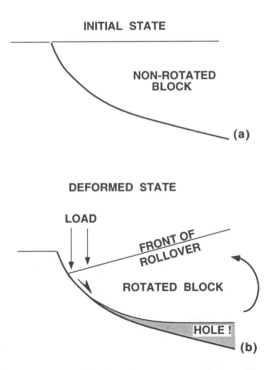

Figure 7-10 Block rotational hypothesis of rollover formation. (a) Before deformation. (b) After deformation and block rotation, a hole opens up between the hanging wall and footwall that can not exist in a gravity field. As this concept of rollover formation is not based on volume conservation, it is not physically reasonable. If applied to a prospect, this concept can lead to errors in interpretation.

In as much as the block rotation concept does not volume balance, its application to a prospect can only result in a structural interpretation that has a volume loss or gain. Rocks are very weak in extension, and the forces of gravity cause the hanging wall to collapse upon the footwall, closing any hole that opens up between the hanging wall and the footwall (examine Figs. 7-1, 7-2, and 7-10b). This gravity collapse of the hanging wall upon the footwall subjects the hanging wall to *internal deformation*. The deformational processes causing this collapse are fundamental to prospect evaluation, and are described in Tearpock and Bischke (1991), in greater detail by Xiao and Suppe (1992), and briefly in Figures 7-1 and 7-2. This is not to imply that deltaic loading does not drive the hanging wall block forward, but rather that loading by itself can not control rollover shape and development.

The reason why balancing concepts are important to prospect generation is that incorrect concepts result in incorrect interpretations. If the slightest problem is present (such as poor data areas), inquire as to which concept drove the interpretation.

Some of the reasons why this is important to prospect evaluation were presented in the previous section on axial surfaces. In addition, interpretation errors tend to be cumulative and may signify that other problems exist. This section concentrates on isolating fault interpretation errors. Let's review several QLTs that can be used to predict fault shape and position, in both good and no data areas.

COMMON FOLD PATTERNS

There are five common fold patterns or styles in extensional regions that can be used to predict fault shape and position. These patterns are:

1. rolled up beds at the front of a rollover;
2. half-graben or monoclinal rollover structure;
3. fanning of bed dips;
4. basinward closure pattern; and
5. the antilistric fault bend.

The first four patterns are discussed next. The antilistric fault bend pattern was discussed earlier in the chapter.

Rolled Up Beds at the Front of a Rollover

The first step in isolating fault interpretation errors is to locate the dip domains and their bounding axial surfaces. As axial surfaces are generated at bends in fault surfaces, dip domain analysis will allow you to predict fault shape and position (Tearpock and Bischke 1991; Xiao and Suppe 1992). Axial surfaces are best located by isolating individual dip domains on profiles (see section on Axial Surfaces Chapter 6). This is done by first locating a zone of near parallel dips and by moving the rolling ruler across the zone of constantly dipping beds. Next, mark the position where the beds change dip on the profile (see Figs. 7-9a and 7-9b). The two adjacent bed dip panels dip at different angles, typically by more than several degrees of dip (Figs. 7-9c and 7-9d). If relative to the regional dip the profile contains *more than one bed dip,* then the causative fault that created the fold has *more than one bend*. For example, fanned bed dips form above *listric fault bends* that contain many axial surfaces (Xiao and Suppe 1992). The seismic sections shown in Figures 7-3 and 7-4 image fanned dips above a listric fault.

When examining a profile such as Figure 7-11a, first locate the shallow beds that exist next to the major expansion fault. These beds dip at gentle angles on the profile. Use dip domain analysis to locate the axial surface that separates the flat dipping beds from the turned up or rolled over beds. Project the axial surface down toward the major fault (Fig. 7-11a). This axial surface defines where the fault changes dip and becomes listric.

Next project the fault surface to the axial surface. The point where the fault surface intersects the axial surface defines where the fault changes dip and becomes more listric (Fig. 7-11b). Fold shape reflects fault shape, so, if the beds do not change dip, then the fault does not change dip (Fig. 7-11a). In Figure 7-11b, there is a simple relationship between fault shape and fold shape. The fault bends up where the beds turn up.

Another example of this relationship is imaged on the seismic section shown in Figure 7-12. In this well imaged example, the fault surface, at first, has a constant dip beneath the flat beds that are next to the fault surface. Then the fault bends at an axial surface, shown as a white dashed line. This axial surface caused the beds to rollover. Notice how the beds turn up to the right of the axial surface.

We have just shown that if a fault is not imaged at depth, the approximate shape and position of the fault surface below the beds can be predicted by analyzing changes in bed dip. We have discussed this process in a qualitative sense. Bischke (1990) and Tearpock and Bischke (1991) present methods that can be used to quantitatively predict fault shape and position. Quantitative prediction methods are beyond the scope of this book.

PATTERN

FLAT BEDS ADJACENT TO MAJOR FAULT SURFACE ARE ROLLED OVER

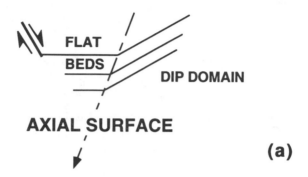

(a)

INTERPRETATION

SINGLE BEND IN FAULT SURFACE

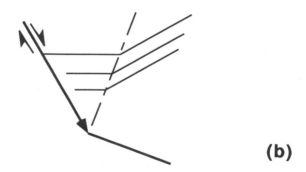

(b)

Figure 7-11 (a) Fault and bed dips at front of a rollover can be used to constrain fault shape and position. (b) Known fault dip is projected to axial surface, where the fault becomes flatter or more listric. Fault shape is related to fold shape, and, thus the fault turns up where the beds turn up.

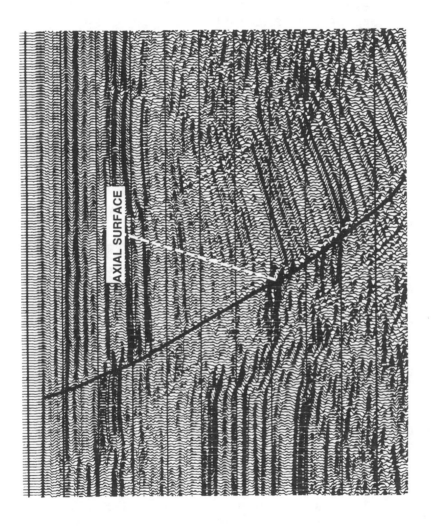

Figure 7-12 Seismic example of relationships described in Figure 7-11. Beds turn up along axial surface which emanates from bend in fault surface. Half-Graben or Monoclinal Rollover Structure. (Seismic published by permission of JEBCO Seismic, Inc., Houston, Texas. Interpretation by Richard Bischke.)

Half-Graben or Monoclinal Rollover Structure

If the rollover is a half-graben structure and contains a relatively flat top (Fig. 7-13), then use the following method to predict fault shape. Using a parallel glider, locate the axial surface positioned between the intersection of the flat and dipping beds at the front of the rollover as shown in Figure 7-13a. In Figure 7-13a, this axial surface is located to the right of the fault. The known fault surface is projected to the axial surface where it changes dip and becomes flatter. If the beds across the crest of the structure are **flat**, then the fault surface does not bend beneath the flat crestal beds. The fault is flat or straight from the point where it intersects the axial surface and can be projected forward at a constant dip beneath the dipping and flat crestal beds (Fig. 7-13b).

The geometric relationships shown in Figure 7-13b can be used to determine the width of the constant dip portion of the major fault and to properly interpret seismic sections. Using Figure 7-14, align a parallel glider along an antithetic fault. Slide the glider across the flat crestal beds and toward the front of the rollover to the position where a *shallow marker bed noticeably changes dip*. The front of the rollover is located next to the major expansion fault. The shallow marker bed must be observed to cross over the top of the rollover and to noticeably change dip. At the top of the rollover where the marker bed changes dip, draw a line to depth that parallels the antithetic fault dip. The marker bed is labelled in Figure 7-14, and the line, or axial surface, is shown as a white arrow dipping to the left.

Repeat this same process at the *back of the rollover* (to the right in Figure 7-14) using a synthetic fault and the same marker bed. Align the glider with a synthetic fault, and move it across the flat part of the marker bed to the back of the rollover where the marker bed changes dip. Next, project a line downward that parallels the synthetic fault dip. This line will dip in the synthetic direction toward the basin (Figure 7-14). The flatter portion of the fault lies roughly between these two diverging white lines (shown in Figure 7-14) that are drawn at the front and back of the rollover. These lines are defined by the rolled-over marker bed and the antithetic and synthetic fault dips. With less coherent reflection data this technique may help you interpret the position of the major fault surface, and will thus allow you to distinguish hanging wall beds from footwall beds.

PATTERN

MONOCLINAL ROLLOVER

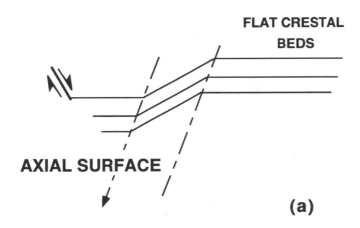

FLAT CRESTAL BEDS

AXIAL SURFACE

(a)

INTERPRETATION

FAULT IS FLAT BENEATH CRESTAL BEDS

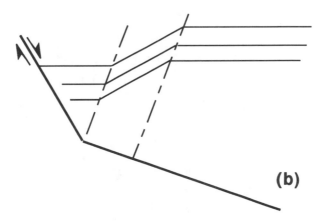

(b)

Figure 7-13 (a) Monoclinal rollover pattern of flat beds at front of rollover, changing to dipping bed on rollover flank. Flat dipping beds exist at rollover crest. (b) Fault can be projected to first axial surface were it turns up, mimicking bed dips. Fault remains flat or straight beneath shallow dipping near surface crestal beds.

194

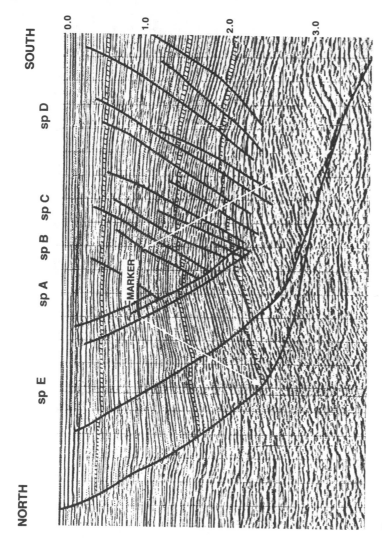

Figure 7-14 Method for determining deep fault geometry. Align rolling ruler parallel to synthetic and antithetic faults, and position the rolling ruler where a shallow marker bed noticeably changes dip. Fault surface will have a flatter dip between projections, defined by marker bed and fault projections. Notice how fault projections locate large bends in the fault surface. (Seismic published by permission of JEBCO Seismic, Inc., Houston, Texas. Interpretation by Richard Bischke.)

Earlier in the chapter, we stressed the importance of being able to recognize prospect location from the shape of major fault surfaces. The relationship between fold shape and fault shape should aid in this analysis. Although the Quick Look Techniques presented in this section are qualitative, in poor or no data areas, these QLTs are better than artistically drawing in a fault on a seismic section. Quantitative techniques for detailed analysis exist to more accurately predict fault position and are presented in Tearpock and Bischke (1991) and Xiao and Suppe (1992).

Fanning of Bed Dips

The next pattern to be described is very common across expansion faults and is widely recognized by geoscientists. If the beds at the front of a rollover have a fan shaped dip pattern or increase their dip with increasing depth, then the fault surface must correspond to the changing dips. This divergent bed dip pattern indicates that the associated major fault is progressively flattening beneath the rolled over beds (Fig. 7-15). This structural style is the most common dip domain pattern observed at the front of rollovers.

Examples of this fanning of bed dips are shown in Figures 7-3 and 7-4. Fanning bed dips imply that an axial surface is associated with each change in bed dip, and that many axial surfaces exist along the listric fault surface (Xiao and Suppe, 1992).

Basinward Closure

Many rollover structures have down-turned beds along their basinward flanks, which dip at angles that are higher than regional dip (compare Fig. 7-3 with Fig. 7-4). These higher dips are often responsible for closing major rollovers in the basinward direction, creating a closure (Fig. 7-3). In this example, Figures 7-3 or 7-4, the major fault also turns down, along its intersection with a **back limb** axial surface. Therefore, if a prospect shows back limb closure but the major fault does not turn down, then the closure may not be real. The axial surface that defines where the fault turns downward emanates from a *convex bend* in the major fault surface (Figs. 7-14 and 7-16). The point at which the fault turns downward can be located by using the dip domain method described earlier. In other instances, a large counter regional (antithetic) fault may close the rollover; care should be exercised when interpreting the cause of the basinward closure.

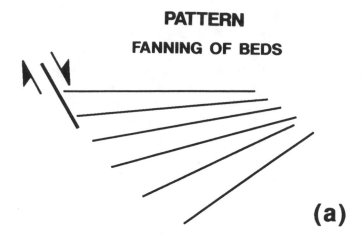

PATTERN

FANNING OF BEDS

(a)

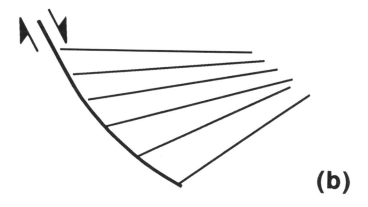

INTERPRETATION

CURVED OR LISTRIC FAULT

(b)

Figure 7-15 (a) Fanning bed dip pattern suggests many minor changes in fault dip resulting in, (b) the common listric normal fault.

Review of Methods and Generalizations

This section on fault shape and fold shape may, at first, appear time consuming and some what complex. With a little practice, there are several generalizations that can help you rapidly analyze seismic interpretations and cross sections. Let's review these methods with an example using seismic lines.

1. Using Figure 7-14, examine the section for bed dips. Typically, the *more steeply dipping beds exist at the front of the rollover,* which is often the landward side of the rollover.

2. Locate the individual dip domains and their associated axial surfaces. This will take only a few minutes after a little practice. Use a clear plastic straight edge or parallel glider for this analysis.

3. Apply these simple procedures. Start at the front of the structure, adjacent to the fault that exhibits the **most expansion**. In Figure 7-14, the major fault is associated with the highest bed dips (sp A to E at 0.0 to 2.2 sec) which define the front of the rollover. Notice that where the bed dips exceed several degrees, a large normal fault is present.

4. Now locate the region of flat or nearly flat dipping beds to the left of sp E in Figure 7-14 adjacent to the major fault surface. These flat bed dips exist at shallow depths where the major expansion fault is typically well imaged. The major fault will begin to flatten or become more listric near where the beds begin to turn up; i.e., fault shape mimics fold shape. The relationship between fold shape and fault shape is shown on Figure 7-12.

5. If the beds are flat across the top (crest) of the rollover, as seen in Figure 7-14 between sp A and B, then the fault is straight at depth, beneath the shallow crestal portions of the rollover (Figs. 7-3, 7-4, and 7-13).

6. If the beds turn down at the back of the rollover, as shown in Figures 7-3, 7-14, and 7-16, the fault also turns down and contains a convex upward bend at depth.

These procedures are simple to use after a little practice. Properly locating faults is certainly worth the effort to save the cost of a dry hole. This will become evident after reviewing the following case study.

PATTERN

ROLL-DOWN

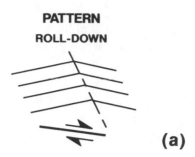

(a)

INTERPRETATION

DOWNWARD BEND IN FAULT SURFACE

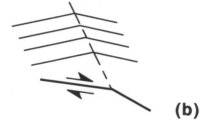

(b)

Figure 7-16 (a) Many rollovers close on the basinward limb of the structure at an angle that is larger than permitted by compaction. (b) Project known fault surface to back limb axial surface. Fault turns down where bed dips turn down.

Case History

Several large discoveries were made in the area shown on Profile A in Figure 7-17a. This profile is well imaged and includes a major listric fault, along with a large synthetic fault (Fig. 7-17a). This geometry, shown in Figure 7-17a, is commonly observed in many extensional environments. After initial success on Profile A, the decision was made to drill the area shown on Profile B (Fig. 7-17b).

Profile B was not, however, as well imaged as A - particularly at the deeper or discovery horizon levels. Furthermore, the large synthetic fault, which was clearly imaged on Profile A, was problematic on Profile B. On Profile B, this synthetic fault was not imaged and may have broken into several faults during its complex history. In addition, Profile B lacked reflector continuity on the deeper levels. The coherent reflections present on Profile B were correlated to Profile A, although the reflections present at the prospect level *could not* be confidently correlated between the two profiles.

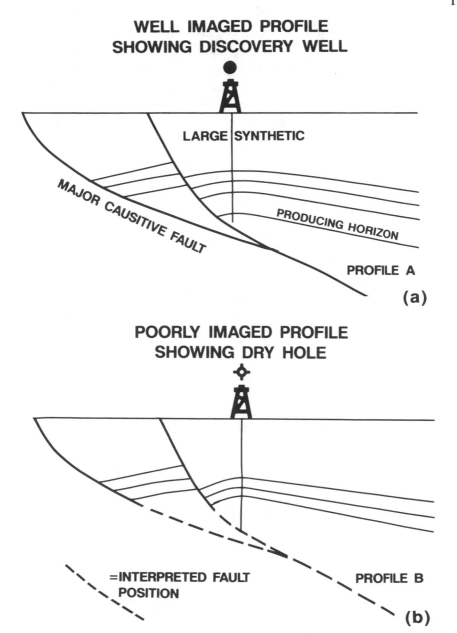

Figure 7-17 Case history of a dry well. Several successful discoveries were drilled along well imaged profile (a), whereas more poorly imaged profile (b) were not geologically successful. Dry wells could have been predicted from change in fold shape imaged on Profiles A and B.

Using the available data and interpretation, a dry hole was drilled on the structure shown on Profile B. This dry hole intersected a stratigraphic section that was described as confusing, and contained formations that were not of the same age as the known hanging wall beds. What went wrong? Could the dry hole have been prevented using the methods described in this section?

POORLY IMAGED PROFILE SHOWING
LOCATIONS OF DRY HOLE
AND SUCCESSFUL WELL

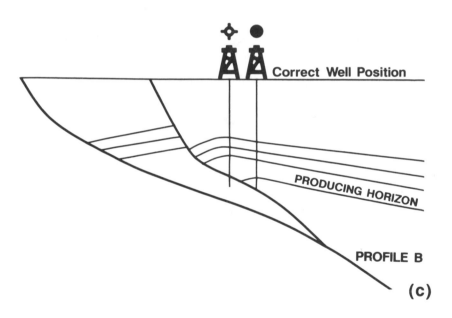

Figure 7-17c Successful well drilled into the hanging wall is positioned to the right of the dry well which penetrated the footwall.

A close examination of Profiles A and B shows that **rollover geometry changed noticeably** between the two profiles. On Profile A, the beds are gently rolled over, whereas on Profile B the beds are **more steeply rolled over.** This change in rollover geometry results from an unrecognized change in fault geometry between Profiles A and B. The beds are gently rolled over on Profile A, and the major fault is primarily responsible for rollover geometry. Consequently, on Profile A, the large synthetic fault did not effect rollover geometry. On Profile B, however, in proximity to the rollover crest, **both frontal and back limb beds dip at a higher angle** than seen on Profile A. The difference in geometry is particularly obvious at the front of the rollover. This change in the dip of the beds above the synthetic fault is caused by changes in the shape of the synthetic fault *rather than* the major fault (compare Fig. 7-17a to Fig. 7-17b).

It follows from the principles discussed in this section that, since the beds are more steeply rolled over on Profile B than on Profile A, the large **synthetic fault is more listric** on Profile B than it is on Profile A. This means that on Profile B, the synthetic fault is on a **higher structural level** (shallower) than it is on Profile A (Fig. 7-17c). After flattening on a higher structural level, the synthetic fault then turns down in order to join the major fault. This has the effect of **turning the basinward flank** of the rollover down, causing the structure to be more tightly folded near the crest. Compare Figures 7-17a to 7-17c. On Profile B, the large synthetic fault, and not the major fault, controlled the position of the prospect. The successful discovery well is located down dip. The reasons for positioning the well in the down dip location are presented in Tearpock and Bischke (1991) and in Xiao and Suppe (1992).

GROWTH QLTs

As large hydrocarbon accumulations are often found within major depocenters subject to expansion faulting (Curtis 1984), we have repeatedly stressed the importance of growth. Although growth is important worldwide, Fisher and McGowen (1967) have shown that most of the production from the Wilcox Group of Texas comes from high growth sections that characterize the delta front. Pacht, et al (1992) present production figures which show that much Gulf of Mexico production is related to the high growth prograding wedge and slope fan system tracts. This has lead to the concept of the production trend, which migrates basinward with time. Growth is important to any area that is subject to expansion faulting. Rollover structures do not exist prior to growth on their causative expansion faults. Therefore, if migration occurred prior to expansion or contemporaneous faulting, then no structure existed to trap the hydrocarbons.

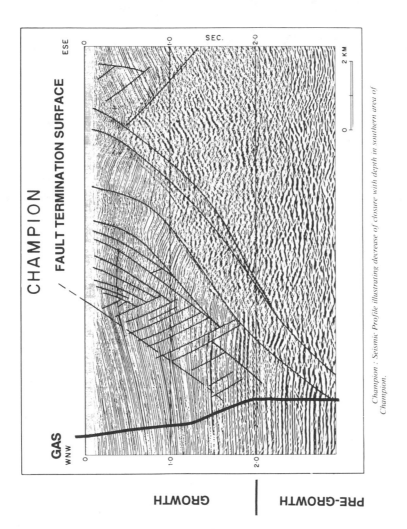

Champion : Seismic Profile illustrating decrease of closure with depth in southern area of Champion.

Figure 7-18 Complex structure from Champion Field, Brunei. Back limb axial surface can be determined using dip domain analysis. Notice that the fault surface turns down where it intersects the axial surface. Unfaulted reflectors that exist adjacent to main fault surface below 2.0 sec indicate pre-growth sediments. (Published by permission of Muzium Brunei.)

Using Antithetic Fault Terminations To Determine Growth

We next introduce several growth QLTs that should enhance your prospect evaluation skills. The first QLT is based on interpreting growth axial surfaces associated with the termination of growth antithetic faults. The theory behind the method is presented in Tearpock and Bischke (1991) and Xiao and Suppe (1992). Antithetic faults typically exist on the crest of rollovers and tend to deform the structure (Figs. 7-3, 7-4, and 7-18). At depth, seismic data loose resolution, and it may not be possible to clearly resolve these antithetic faults. We can, however, predict their presence, as faulting tends to make reflections **non-coherent** below 3.0 sec levels (Fig. 7-19a). Notice that, unlike major expansion faults, crestal antithetic faults **age** with depth into the basin (Figs. 7-18 and 7-19a). We use this aging of the antithetic faults to define antithetic fault growth.

Figure 7-18 images a complex structure subject to both extension and compression. (The structure is an inversion structure.) However, the growth stage of the extensional deformation is still evident. In Figure 7-18, the upper terminations of the antithetic faults occur along a line that plunges into the basin. This line indicates where the faults **cease to grow** upward and are buried by younger growth sediments. This line can be used to determine growth, or, more precisely, when growth faulting stopped. Draw a line along or tangent to the uppermost termination of the antithetic faults as they die off into the basin. In Figure 7-18, growth faulting, and hence growth exists only above about 2.0 sec. There is other evidence that growth occurs above this level. The intervals above 2.0 sec are expanded, whereas the intervals below 2.0 sec are not (see section on Expanded Reflection later in this chapter). Further evidence for no growth below 2.0 sec is supported by unfaulted, coherent reflectors that intersect and exist next to the major fault surface. This means that the section below 2.0 sec is unaffected by the growth faulting, and is a pre-growth section. This section was not rolled over at the time these beds were deposited. If migration occurred prior to this time, then no structure was present to trap hydrocarbons.

At greater depths, where seismic data looses its ability to resolve the antithetic faults, the faulted section can be identified by mapping the boundary between the coherent and non-coherent reflectors. This is shown in Figure 7-19b as a dashed line. This interpretation of the non-coherent data assumes that the growth faulting makes the reflectors non-coherent. Tearpock and Bischke (1991) show that as long as this growth fault termination surface does not intersect the major fault surface, or coherent reflectors do not intersect the major fault surface, then the sedimentary section is a **growth section.**

Using Axial Surfaces to Determine Growth

In this section, we present a **growth axial surface** technique that defines the higher growth sections. Most geologists know that, in the extensional growth fault areas, large hydrocarbon accumulations tend to exist in thick depocenters associated with growth or expansion faulting, rather than in low growth or pre-growth sections (Fisher and McGowen 1967; Curtis 1984; Pacht, et al. 1992.)

Growth axial surfaces (GAS) can be used to categorize growth on rollovers. However, growth axial surfaces are often subtle features, and good data are required to identify them. At times, crestal growth faulting tends to **mask and obscure** axial surfaces associated with rollovers. However, if the crestal faulting is minimal or absent, then axial surfaces can be used to determine growth.

When mapping axial surfaces we must locate the dip domains. First, locate the dip domains that are at the back or basinward side of the rollover. The growth axial surface separates back limb bed dips from the more gently dipping basinal beds (Fig. 7-18). The growth axial surface that we are attempting to locate is at the break in slope (Fig. 7-18) between the steeper dipping beds on the flank of the rollover, and the more gently dipping beds in the basin. In Figure 7-18 growth initiates where the GAS terminates at a pre-growth axial surface that dips at a higher angle below 2.0 sec.

If the back limb axial surface is not easily recognized because of crestal faulting, or if the rollover is a half-graben or monoclinal structure (Fig. 7-13), then locate the GAS that is positioned near the front of the rollover. This front limb axial surface may be easier to locate than the back limb axial surface. Using dip domain analysis, locate the region of flatter bed dips that is between the **crest** of the rollover and the more steeply dipping beds located at the front of the rollover.

An example of analyzing axial surfaces is shown in Figure 7-20, from Slick Ranch Field, Texas. Although the Slick Field is a complex example, growth axial surfaces can be readily located. A parallel glider was first passed across the data to locate the region of nearly flat dipping beds that define the crest of the rollover (bold lines in Fig. 7-20). Figure 7-20 contains several major unconformities and regional tilting, so the ruler was repositioned to account for unconformities and dip changes caused by regional

205

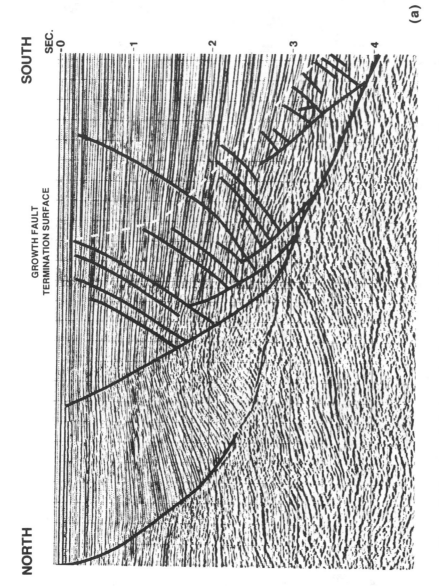

Figure 7-19a Growth fault termination surface (dashed line) above a major listric normal fault (solid line). Growth fault termination surface is mapped at the upward termination of antithetic faults or where coherent reflectors lose continuity (coherency). (Seismic published by permission of JEBCO Seismic, Inc., Houston, Texas. Interpretation by Richard Bischke.)

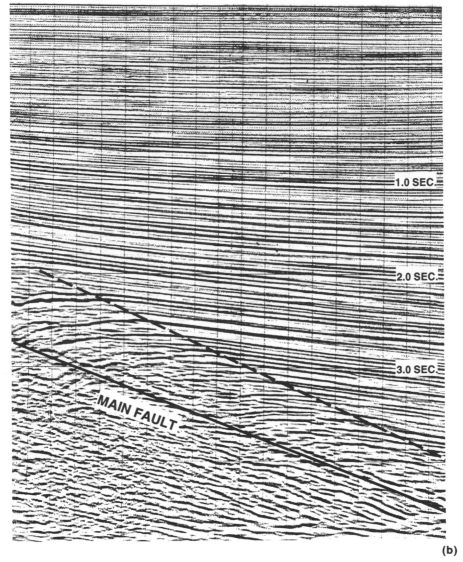

Figure 7-19b Non-coherent deep reflectors are interpreted to indicate growth section. (Seismic published by permission of JEBCO Seismic, Inc., Houston, Texas. Interpretation by Richard Bischke.)

tilting. Notice that a fan dip pattern is located at the front of the structure. The growth axial surface is defined by the change in bed dips that occurs between the crestal and frontal dip domains (dashed line in Fig. 20). Again, as long as the axial surface does not intersect the main fault surface, the section is a growth section.

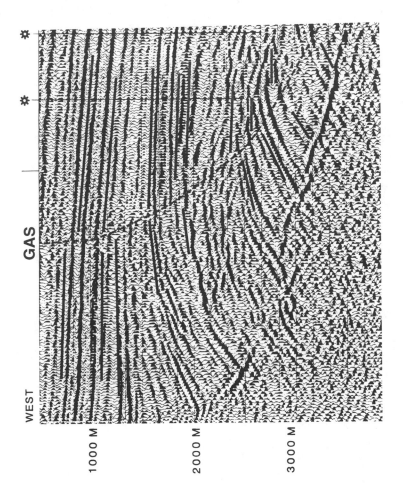

Figure 7-20 Complex example of front limb growth axial surface in Slick Ranch Field, Texas. Rollover is subject to regional tilting and several major unconformities. Growth axial surface (dashed line) is defined by changing bed dips that exists between frontal and crestal beds (bold lines). Production comes from section containing a shallower or more gently dipping growth axial surface. (Published by permission of the American Association of Petroleum Geologists. Interpretation by Richard Bischke.)

208

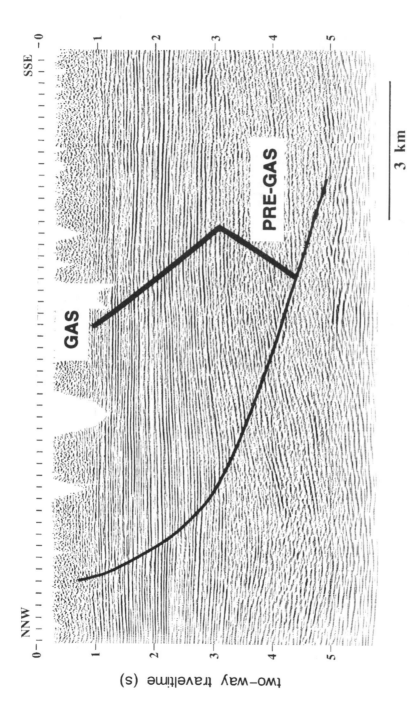

Figure 7-21 Growth axial surface terminates at a pre-growth axial surface at just below 3.0 sec. Structure did not exist at the age equivalent of 3.0 sec; thus, section above about 3.0 sec is a growth section. Rolling ruler can be used to locate subtle changes in bed dip. (From Xiao and Suppe (1992). (Seismic published courtesy of Texaco, interpretation by Richard Bischke.)

In extensional areas, growth axial surfaces may terminate against a pre-growth axial surface. The intersection of the growth axial surface with the pre-growth axial surface defines the initiation of growth. Thus, the fold began to grow at the point where the growth axial surface terminates or changes direction. A rollover from southern Louisiana, shown in Figure 7-21, began to grow rapidly above 3.0 sec. In this case, growth axial surface analysis might be critical to prospect evaluation of a deep prospect, since a structural trap did **not exist** when units corresponding to reflections below 3.0 sec were deposited.

Slope of Axial Surface as a Potential Hydrocarbon Indicator

Our empirical observations of the slopes of growth axial surfaces, taken on rollovers from various portions of the world, suggest that *major production* often correlates to sections where the *slope of the growth axial surface is low*. Two examples of this observation are shown on a rollover from Brunei (Fig. 7-22) and on a rollover from the United States Gulf Coast (Fig. 7-25). In the Brunei example, production comes from the stratigraphic sections where the slope of growth fault termination surfaces is low or where growth is high. Another example, from Texas, is shown in Figure 7-20 (below 2400 M), where production comes from levels where the growth axial surface has a low slope.

We suggest that these relationships may be significant, as hydrocarbons probably migrate along fault surfaces (Weber and Daukora 1976; Price 1980; Leach 1993). The migrating hydrocarbons should have the effect of increasing pore pressures on these surfaces, thus buoying the fault surfaces (Hubbert and Rubey 1959). This buoying or weakening of the fault surface should increase the rate of fault slip. The higher the pore pressures possibly created by migrating hydrocarbons (Leach 1993), the less the effective friction; thus, the hanging wall slips faster over the footwall.

In Figure 7-23, Xiao and Suppe (1992) have generated several generic rollover models at varying sedimentation to fault slip rates. In this figure, the sedimentation rate is plotted on the y-axis and the fault slip rate on the x-axis. Notice when the sedimentation rate is high relative to the fault slip rate, the growth axial surfaces have a **steep slope** (Figs. 7-23a and 7-23c), and the sedimentary packages are thin (Figs. 7-23a, 7-23c, and 7-23e). In contrast, when the fault slip rate is high relative to the sedimentation rate, the growth axial surfaces have a **lower slope** and the sedimentary **packages are thicker**

210

or expanded (Figs. 7-23b, 7-23d, and 7-23f). We suggest that hydrocarbon accumulation may correlate with the intervals where growth axial surfaces have a lower, rather than a higher slope. This idea can be used to define the more prospective section (Figs. 7-20, 7-22, and 7-25) when exploring for rollover structures.

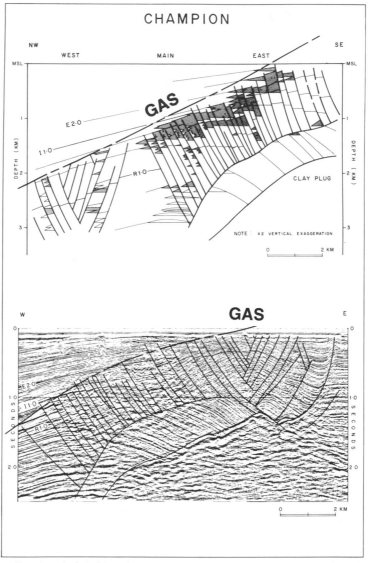

Champion : Geological Cross Section and Seismic Profile.

Figure 7-22 Shallow dipping growth fault termination surface correlates with major production. Below about 2.2 sec crestal faulting stops and coherent reflectors intersect main fault surface, indicating a pre-growth section. (Published by permission of Muzium Brunei.)

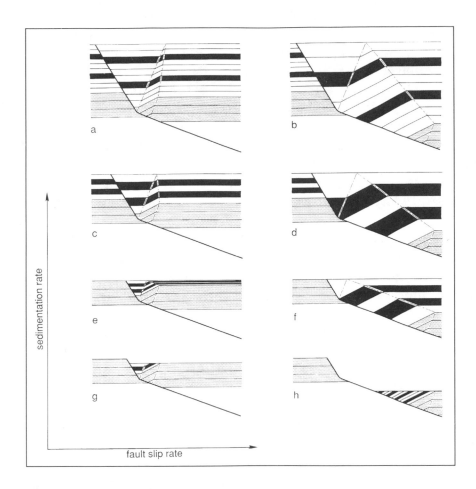

Figure 7-23 Dips of growth axial surface relative to sedimentation/fault slip rate. Where sedimentation rate is high and fault slip rate is low the growth axial surface has a steep slope (a and c). These sections are generally less productive. If fault slip rate is high relative to the sedimentation rate (b, d, and f), then growth axial surface has a more gentle slope. These expanded sections (b, d, and f) may correlate to rapid fault slip rates, thick sand sections and the hydrocarbon migration stage. (From Xiao and Suppe (1992). Published by permission of American Association of Petroleum Geologists.)

EXPANDED REFLECTIONS:
HOW TO IDENTIFY DEEPER GROWTH ZONES

In areas where well control is lacking, locating sands across *major* expansion faults is often problematic. One approach is to use the antilistric shape of crestal antithetic faults to locate and estimate percent sand. However, due to the depth resolution on seismic sections, this is not always possible below depths of 15,000 to 20,000 feet.

Another approach is to locate *expanded sedimentary sections* (Figs. 7-23d and 7-23f) on seismic lines. Highly expanded sections often image as low frequency reflections at depths exceeding 15,000 feet. One can apply the following technique to locate expanded sections where well control or antilistric faults are lacking. Examine the reflection pattern of the *boldest* reflectors across a deep rollover structure, such as the one illustrated in Figure 7-24. The bolder reflectors often represent *sequence boundaries or boundaries of larger sedimentary packages.* In Figure 7-24, measure with a scale, or in the mind's eye, the thickness between the bold reflectors located near the front (T_2) and at the crest (T_1) of the rollover. Notice that the ratio of T_6/T_5 is near unity, but that the ratios of T_2/T_1 and T_4/T_3 are greater than unity. Therefore, we can conclude that the greater the ratio, the more likely that the interval has been expanded by normal faulting, and the greater the likelihood that the interval contains thicker sand formations (Thorsen 1963; Fisher and McGowen 1967). However, growth or expansion faulting can not guarantee the presence of sands.

Figure 7-23 shows 8 different models that contain various amounts of growth. Growth is low where the fault slip rate is low and the sedimentation rate is high (Figs. 7-23a, 7-23c, and 7-23e). High growth rate models are shown in Figure 7-23b, 7-23d, and 7-23f, where the fault slip rate is high and where the sections are highly expanded. Again, high growth sections often contain the greatest hydrocarbon potential. Another example of a highly expanded and productive section is the Big Hum, Corsair Trend (Vogler and Robinson 1987) shown in Figure 7-25. Notice how the reflections between 3000 and 5000 m expand and diverge toward the major Corsair Fault.

EXPANSION INDICES AND ΔD/D DIAGRAMS (a 30 Minute Quick Look Technique)

The Expansion Index Technique is often applied across expansion, growth, or syndepositional faults to determine the beginning of growth and to identify the higher growth intervals. These high growth intervals often contain thick reservoir sands (Thorsen 1963; Tearpock and Bischke 1991). In previous sections, we outlined how rollovers form over major growth faults. Consequently, in deltaic areas, we are confronted with the ideal situation in which thick reservoir units are contemporaneously folded into structural traps.

LOCATING DEEP EXPANDED SECTIONS

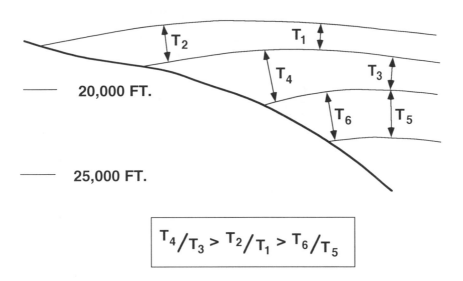

$$T_4/T_3 > T_2/T_1 > T_6/T_5$$

Figure 7-24 Expanded sections often correlate to thick reservoir quality rocks. The extent to which a section is expanded can be roughly estimated by comparing frontal limb sequence thicknesses to crestal sequence thicknesses.

The Thorsen Expansion Index is calculated from well logs or depth corrected seismic data, by measuring stratigraphic thicknesses downthrown, and comparing it to the equivalent upthrown thicknesses. The higher the expansion index, the higher the growth, or

$$\text{Expansion} = \frac{\text{Thickness Downthrown}}{\text{Thickness Upthrown}}$$

This technique has the disadvantage of averaging across disconformities or hiatuses.

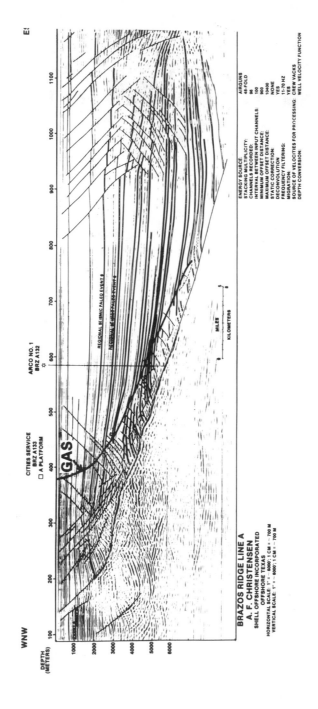

Figure 7-25 Productive Brazos Ridge or Big Hum trend, offshore Texas. Notice shallow dipping growth fault termination surface and expanded section at ARCO No. 1 BRZ A132 Well. (From Christensen, in Bally (1983). Published by permission of American Association of Petroleum Geologists.)

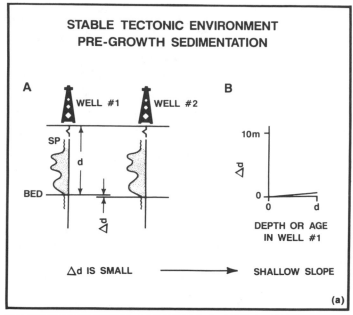

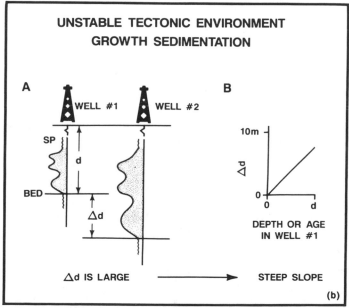

Figure 7-26 Powerful, but rapidly applied, Δd/d technique. (a) The offset in beds (Δd) are plotted relative to corresponding beds in structurally higher Well No. 1 (d), then a growth diagram is constructed. If the slopes on the diagram are shallow or flat, then the section is a pre-growth section. (b) Repeating this process in a growth environment results in steep slopes on the Δd/d diagrams. Method can be used to identify a variety of structural-stratigraphic problems.

Bischke and Elston (1991) have developed a similar, but more accurate technique with a *higher resolution* based upon distances that parasequence boundaries or shale breaks are vertically displaced. This is the distance that the beds have been vertically displaced during growth faulting. The technique uses well log or seismic data to correlate beds or shale breaks across an expansion fault. Using well logs, two dip oriented wells are required for the analysis. Once the well correlations are made, the vertical distances that the horizons are *vertically displaced* (Δd) are plotted against the total depths in the *updip or structurally higher well* (d). This technique is illustrated in Figure 7-26. The results are plotted in a Δd/d diagram, where Δd is plotted on the y-axis and d is plotted on the x-axis. On the Δd/d or differential growth plots, the **higher the slope on the plot, the higher the differential growth** (Fig. 7-26b). Pre-growth sections reflect a *shallow or flat* slope, as shown in Figure 7-26a.

Bischke and Elston (1991) and Bischke (1994) have shown that, when applied to *depth corrected* seismic sections taken from the front of rollovers, the Δd/d method, can be used to rapidly:

1. identify the initiation of growth;
2. locate subtle sequence boundaries and subtle stratigraphic traps **or disconformities;**
3. locate data miscorrelations;
4. identify expanded and condensed sections;
5. help in distinguishing faults from unconformities; and
6. identify a variety of stratigraphic and structural problems that are not readily recognized using conventional interpretation techniques.

Figure 7-27 illustrates the method, using seismic correlations that are depth corrected. In this area, sequence boundaries occur as disconformities. Common interval velocities were used to depth correct the seismic data. Often, the growth data are **roughly linear over extended periods of time,** and then change to another period of linear, but changing growth. A change in growth is likely to occur across a hiatus. A change in sedimentation or sea level is expressed as a change in slope on the plot. Such changes are important. In looking at Figure 7-27, we can conclude that sequence boundaries are located near breaks in slope on the curves, or at Correlation Point Nos. 8, 12, 18, and 22.

Also notice on Figure 7-27, between Correlation Point Nos. 12 and 22, that a high growth section exists from the 2500-ft to the 5000-ft levels (refer again to Fig. 7-26b). From well log analysis, good reservoir sands are present between Correlation Point Nos. 18 and 22. This interval averages over 50% sand. Furthermore, the technique identifies the *condensed shale section* located below Correlation Point No. 22. Condensed sections, which are sometimes source rock sections (Vail and Wornardt 1991), occur as *negative slopes* on Δd/d plots.

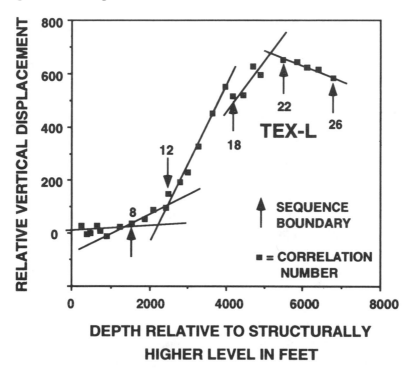

Figure 7-27 Depth corrected Brazos Ridge Δd/d plot, showing subtle sequence boundaries at Correlation Point Nos. 8, 12, 18, and 22. Unconformities or disconformities are defined by break in slope, or at discontinuities in the curve. Notice highly expanded section at Correlation Point Nos. 12 to 22. Percent sand averages 50% from No. 18 to No. 22. Negative slope below No. 22 indicates a condensed shale section.

Assume that Figure 7-27 is generated from depth corrected seismic data taken from a frontier basin, and that a well is planned to test the section down to Horizon No. 26 at 7000 ft. The plot can be used to predict the following: (1) the section from Correlation Point Nos. 1 to 8 is a low or pre-growth section and may not be as prospective, (2) slow growth ends at Correlation Point Nos. 8 to 12, but accelerates between Correlation Point Nos. 12 to 22, with thick reservoir quality sands most likely existing between these correlation numbers, and (3) the section below Correlation Point No. 22 is probably a shale section. So, if it is not your intent to test for source rocks, the well can test the primary section to a depth limit of about 5000 ft. In addition, hydrocarbon potential is greatest between 2,500 and 5,000 ft and is poorer above 2,500 ft and below 5,000 ft.

The Expansion Index and the $\Delta d/d$ methods represent techniques that can be rapidly applied. Prior to a presentation, you can consult two correlated well logs, extract their depths, rapidly enter the data onto a spreadsheet, and display the data in a graphics program. Subtract the depths in the structurally lower well from the depths in the structurally higher well. These values are Δd and are plotted on the y-axis. Then plot Δd against the correlative depth in the structurally higher well. For best results, the two wells should be aligned in the **dip direction,** and the parasequence correlations are best taken at 100 to 200-ft intervals. This simplifies the interpretation of the diagrams. As a QLT, a manager or supervisor can ask the simple question, "Have we applied any growth analysis such as $\Delta d/d$ over our prospect"? If the answer is "no", and growth or the presence of expanded section is important, then growth analysis can be requested.

These plots represent **information that is present in the well log data,** and thus any **interpretation of the area must be consistent with the correlations as they are displayed on these plots.** It also follows that an interpretation under review that is **inconsistent with these plots cannot be correct** (Bischke and Elston 1991; Bischke 1994). Remember, growth records the details and subtleties within the stratigraphic section, which may provide insight into potential stratigraphic and structural problems inherent in the prospect area. We find that the techniques can be used to rapidly identify correlation busts and to confirm small faults.

After analyzing a $\Delta d/d$ plot from a specific prospect, you are in a position to ask questions regarding the prospect and potential problems. Such questions may include: (1) is there structural growth at the level of interest, (2) are there potential source (condensed) and reservoir (highly expanded) sections present, and (3) are there any potential stratigraphic and structural problems, such as the thinning of sands, stratigraphic traps, late fault timing, or downward dying growth faults. The $\Delta d/d$ technique directs you to potential problem areas and may quickly help resolve these problems.

Finally, we present a summary of how to interpret $\Delta d/d$ patterns. A more detailed explanation of the technique is presented by Bischke (1994).

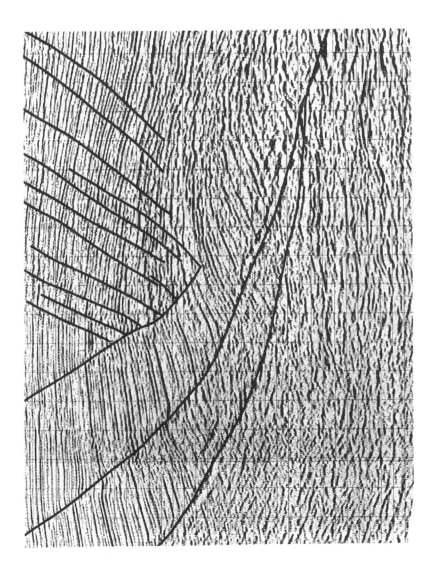

Figure 7-28 Downward dying crestal growth faults terminating above coherent reflectors. (Seismic published by permission of JEBCO Seismic, Inc., Houston, Texas. Interpretation by Richard Bischke)

1. A positive slope indicates an expanded section; the steeper the slope, the greater the growth (Δd plotted on y-axis and total depth, d, on the x-axis). Provided sands are present, high growth sections have a high reservoir rock potential.

2. A negative slope indicates a condensed or potential source rock section. Condensed sections are often correlatable from basin to basin (Vail and Wornardt 1992).

3. Large discontinuities on the diagrams indicate large unconformities or faults. If the discontinuity is caused by an unconformity, then an increasing or positive offset in Δd indicates an onlap, while a decreasing offset indicates a downlap.

4. Disconformities, small unconformities, and subtle sequence boundaries occur as a change in slope.

5. Data miscorrelations plot off of the general trend of the Δd versus d curve.

DOWNWARD DYING GROWTH FAULT PITFALL

Growth faults that **decrease** their vertical separation with depth have been described by Ocamb (1961), Bischke and Suppe (1990), Tearpock and Bischke (1991), Xiao and Suppe (1992), and Bischke and Suppe (1993). These faults affect the very existence of deep plays, and it is important to recognize them. If either a seismic or well log data set indicates that the vertical separation on the fault being studied decreases with increasing depth, then a downward dying growth fault is probably present. On seismic profiles, these faults have offsets on shallower horizons that are larger than offsets on deeper horizons. The slip on downward dying faults may go to **zero**, with the faults terminating above coherent reflectors (Fig. 7-28). In well logs, these faults have less missing section at greater depths.

Downward dying faults are more common than realized, and their presence becomes extremely important when drilling deep prospects. A fault dependent footwall trap on a downward dying growth fault will cease acting as a seal once the vertical separation becomes less than reservoir thickness. Furthermore, if the *fault dies to zero,* the play may not exist where it is mapped (Fig. 7-29a). An alternate play may exist in a structurally higher updip section, such as shown in Figure 7-29b. The alternate prospect may have been previously drilled or may still have potential. We have seen several very expensive dry holes drilled for footwall traps that did not exist.

Some faults do die downward, but if an explorationist does not accept this fact or does not recognize a downward dying fault, then a non-existent prospect, as shown in Figure 7-29b, might be drilled. Such non-existent plays can become costly dry holes. Refer to Tearpock and Bischke (1991) for a discussion of the origin of downward dying growth faults.

CORRECT INTERPRETATION

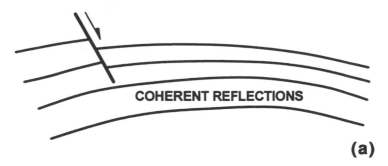

INCORRECT INTERPRETATION

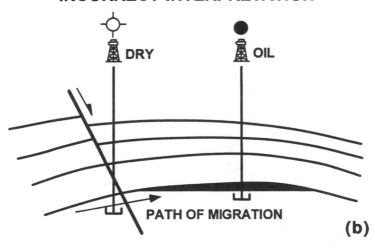

Figure 7-29 Downward dying growth fault pitfall. (a) Correct interpretation, fault dies with depth into coherent reflections. (b) Incorrect interpretation results in a dry well. If a fault is pushed through coherent reflectors, then a non-existent structural trap is created. In this case hydrocarbon migrated to a structurally higher position.

Strike Ramp Pitfall

We do know, however, that if a region has been subject to extension, the sum of the slip on all the normal faults is likely to remain roughly constant across an area (Ramsey 1967), as illustrated in Figure 7-30. Therefore, if a series of en èchelon faults exist in map view (Fig. 7-30, shallow level), the slip or vertical separation along the en èchelon faults at Profile A is roughly equal to the slip on the faults along Profiles B and C. If a map contains en èchelon offset faults and the vertical separation does not have the appearance of adding between adjacent profiles, then the interpretation is likely to be in error. For an extended discussion on transfer zones refer to Morley et al. 1990.

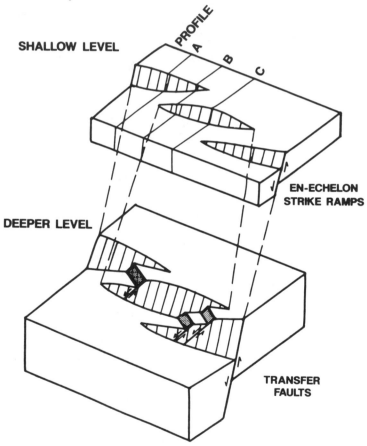

Figure 7-30 Slip consistency across normal faults. Slip must roughly add from one dip profile to the adjacent dip profile. (Modified from Modern Structural Geology, volume II, Ramsey and Huber. Published by permission of John Ramsey.)

Figure 7-30 shows a series of en èchelon faults that, on the shallow level, will not seal laterally. If, for example, an interpreter has Profiles A and C, the interpreter is likely to connect the fault on Profile A to the fault on Profile C. A non-existent trapping fault may be created, particularly on the shallow level of Figure 7-30. En èchelon faults and their associated strike ramps are **common**, and if fault surface maps are not constructed, their presence may not be detected. How do we recognize the presence of a strike ramp or two faults dying in opposite directions along strike? A slight bend along a fault trace on a structure map may indicate the presence of two faults or a strike ramp. If a bend in a fault trace is observed, then, as always, a fault surface map should be constructed.

If a strike ramp is subjected to larger displacements than those shown on the shallow level of Figure 7-30, then the strike ramp may stretch and become subject to normal faulting on a deeper level. The strike ramp may fault through and form a sealing fault that is later than the en èchelon faults shown on the shallow level (Fig. 7-30). Fault timing of the strike ramps is important and the hydrocarbon migration must be later than the normally faulted strike ramps.

In this chapter we have presented a number of extensional QLTs that should help you more critically evaluate prospects, and rapidly locate normal fault pitfalls and locate intervals that have a high potential for hydrocarbons. These QLT techniques include using fault shape to check the shape of hanging wall anticlines on cross sections and on maps, rapidly locating large rollover highs on seismic sections or on fault surface maps, using fault shape as an indication of lithology, using axial surfaces to constrain fault shape and position, using growth faults and growth axial surfaces to locate intervals that have a high potential for hydrocarbons, using expanded reflections to locate high growth intervals, using the high resolution $\Delta d/d$ technique to solve and to locate stratigraphic or structure problems, and lastly we studied the downward dying and strike ramp normal fault pitfalls. Several minutes spent critically reviewing maps and cross sections utilizing these QLTs can save the costs of dry holes and subsequently redoing projects.

CHAPTER EIGHT

DIRECTIONAL WELL
QUICK LOOK TECHNIQUES (QLTs)

INTRODUCTION

Geological interpretations become more difficult with the introduction of directionally drilled wells. Electric logs from directionally drilled wells depict the apparent thickness of the stratigraphic interval with respect to the angle of wellbore deviation and the dip of the beds.

If the formation is not flat, these apparent log thicknesses, if not corrected, are wrong with respect to the size of a fault or the amount of net sand, gas, or oil for a given reservoir. This chapter presents QLTs related to directionally drilled wells.

VERTICAL SEPARATION AND DIRECTIONALLY DRILLED WELLS

Has the correct missing section (vertical separation) by log correlation, been used in the preparation of structure maps? Vertical Separation is defined as the true vertical thickness (TVT) of the stratigraphic interval that has been faulted out (missing section) or repeated (repeated section) in a wellbore when compared to one or more other wellbores that have the same stratigraphic interval thickness (Tearpock and Bischke 1991). **True Vertical Thickness (TVT) is the thickness that is seen in a vertical**

well. Therefore, the estimated missing or repeated section for any fault used to prepare a structure map must be determined by correlating the faulted well with a vertical or straight hole, or correlating the faulted well with a deviated well and *correcting* the missing or repeated section for wellbore deviation and bed dip.

Figure 8-1 is a cross section illustrating a faulted horizon (Bed A) penetrated by one vertical well and two deviated wells. The vertical well cuts a fault that completely faults out Bed A. With the cross section, we can see that the missing section in Well No. A-1 is 200 ft. Considering that the amount of missing section (vertical separation) is defined as the true vertical thickness of the stratigraphic interval faulted out, and that the fault in Well No. A-1 completely faults out Bed A, the cross section shows that the size of the fault cut in Well No. A-1 is equal to the vertical thickness of Bed A, which is 200 ft.

When directionally drilled Well Nos. A-2 and A-3 are used to correlate with Well No. A-1, the missing section obtained is in terms of *deviated* log thickness rather than true vertical log thickness (correlative missing section with Well No. A-2 = 504 ft and with Well No. A-3 = 164 ft). The deviated log thicknesses must be converted to true vertical thickness (TVT) to obtain the true missing section (correct value for vertical separation) for mapping purposes. To calculate the true vertical thickness for the fault cut, the missing section data determined from Well Nos. A-2 and A-3 and a three-dimensional correcting factor equation must be used (Setchel 1954). The following equation can be used to make these corrections.

$$\text{TVT} = \text{MLT} [\text{Cos } \Psi - (\text{Sin } \Psi \text{ Cos } \alpha \text{ Tan } \phi)] \quad \text{Equation 8-1}$$

To calculate a correction factor using the general three-dimensional equation, the required data are: (1) wellbore deviation angle (Ψ) obtained from a directional survey, (2) wellbore deviation azimuth (α_w) obtained from a directional survey, (3) true bed dip (ϕ) measured from a completed structure map or dipmeter, if run in the wellbore, (4) bed dip azimuth (α_b) measured from a completed structure map or obtained from a dipmeter, (5) the delta azimuth (α) which is the difference between the wellbore azimuth (α_w) and the true bed dip azimuth (α_b), and (6) the measured logged thickness (MLT) in the deviated well.

226

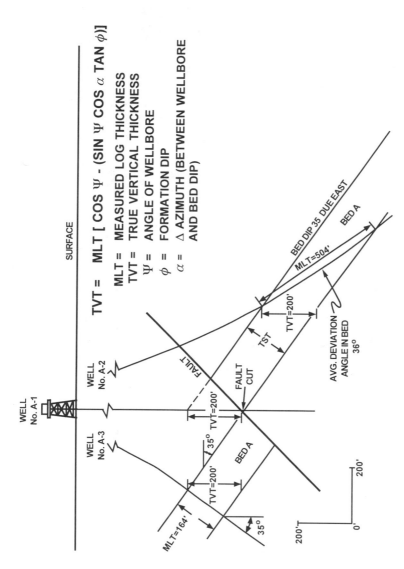

Figure 8-1 One vertical and two deviated wells penetrate Bed A. In Well No. A-1, Bed A is faulted out. The cross section shows the relationship of the missing section in Well No. A-1 to the measured log thicknesses in deviated Well Nos. A-2 and A-3 that are equivalent to the missing section. Equation shown is used to correct measured log thickness to true vertical thickness. (Published by permission of Prentice-Hall, Inc.)

The corrected missing section (TVT) for the fault as correlated with Well No. A-2 is:

Data:

$$\Psi = 36 \text{ deg}$$
$$\alpha_w = 90 \text{ deg}$$
$$\phi = 35 \text{ deg}$$
$$\alpha_b = 90 \text{ deg}$$
$$\text{MLT} = 504 \text{ ft}$$
$$\alpha = 0 \text{ deg, } \Delta \text{ azimuth}$$

$$TVT = MLT [\text{Cos } \Psi - (\text{Sin } \Psi \text{ Cos } \alpha \text{ Tan } \phi)]$$

$$= 504' [\text{Cos } 36° - (\text{Sin } 36° \text{ Cos } 0° \text{ Tan } 35°)]$$

$$= 504' [0.8090 - (0.5878) (1) (0.7002)]$$

$$= 504' [0.8090 - 0.4116]$$

$$TVT = 504' [0.3974]$$

$$\mathbf{TVT = 200 \text{ ft}}$$

The TVT Calculation for the missing section as correlated with Well No. A-3:

Data:

The data for this calculation are exactly the same as for Well No. A-2, with two exceptions. The azimuth (α_w) for Well No. A-3 is due west or 270 deg, therefore the Δ azimuth (α) is 180 deg and the wellbore deviation angle (Ψ) is 35 degrees.

$$TVT = MLT [\text{Cos } \Psi - (\text{Sin } \Psi \text{ Cos } \alpha \text{ Tan } \phi)]$$

$$= 164' [\text{Cos } 36° - (\text{Sin } 35° \text{ Cos } 180° \text{ Tan } 35°)]$$

$$= 164' [0.8192 - (0.5736) (1) (0.7002)]$$

$$= 164' [0.8192 - (-0.4016)]$$

$$= 164' [0.8192 + 0.4016]$$

$$TVT = 164' [1.2208]$$

$$\mathbf{TVT = 200 \text{ ft}}$$

From these calculations, it is determined that the fault in Well No. A-1 has a vertical separation of 200 ft. It is this 200-ft value that must be used in the preparation of the structure maps or cross sections.

When working with moderately deviated wells (less than 20 deg) in areas of horizontal beds or low angle bed dips (less than 5 deg), the method for determining the true vertical thickness of the missing section becomes relatively simple. If the beds are horizontal (Fig. 8-2), the correction reduces to the simple trigonometric solution of a right triangle. Therefore, as shown in Figure 8-2, the true vertical thickness for a deviated well penetrating horizontal beds is determined by multiplying the measured log thickness by the *cosine* of the wellbore deviation angle. Therefore:

$$TVT = MLT (\text{Cos } \Psi) \quad \text{Equation 8-2}$$

This simplified correction factor should be limited to areas with bed dips ≤ 5 deg and wellbore deviation angles of ≤ 20 degrees. Wells with a very high borehole deviation angle and especially, horizontal holes should be corrected for TVT with the entire equation (Eq. 8-1) even when the bed dips are low. Equation 8-2 only corrects for wellbore deviation. In flat beds TVT is equal to TVDT (True Vertical Depth Thickness). In dipping beds TVT is not equal to TVDT.

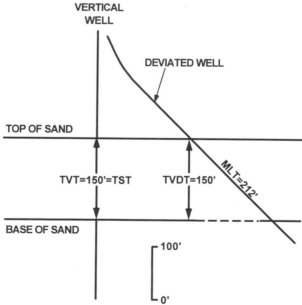

Figure 8-2 The true vertical thickness, true vertical depth thickness, and the true stratigraphic thickness calculated from a deviated well have the same value when the beds are flat (Published by permission of Prentice-Hall, Inc.)

In preparing or evaluating prospect maps associated with a fault and directionally drilled wells, be sure that the missing or repeated section has been corrected back to true vertical thickness before it is used to prepare a structure map. Consider a 1500-ft fault cut, based on correlation with deviated wells, that corrects back to 900 ft. If the correction is not made, the completed structure map will incorrectly incorporate into the interpretation a 1500-ft fault, rather than a 900-ft fault. Such mistakes can have a significant impact on a prospect map.

In summary, when preparing maps in an area with deviated wells and dipping beds, the vertical separation, obtained by correlation with logs from deviated wells, **must** be corrected to true vertical thickness. When evaluating prospects, be sure to verify that these corrections have been made. If they have not been made, the impact of the corrected fault cut on the interpretation should be considered before any investment decisions are made.

PRODUCTION FAULTS OR WATER LEVEL CHANGES - REAL OR THE RESULT OF DIRECTIONAL SURVEY UNCERTAINTIES

Figure 8-3 is a structure map on the top of the 8700-ft Sand. Notice that the gas/water contact in the northern portion of the field is -8766 ft, while to the south it is -8686 ft. This is an 80-ft difference. A production fault to the west and a water level change to the east have been incorporated into the map, in an attempt to explain the difference in water level.

With a quick look at the structure map, one can see that most of the wells in the field are directionally drilled from no less than five different platforms. One should ask, "Is the apparent variation in water level a geologic or a directional survey problem?" Directional surveys have uncertainties associated with them. Wolff and de Wardt (1981) undertook an extensive study of the lateral and vertical uncertainty of various types of directional surveys. The results of their study are generalized in Figures 8-4a and 8-4b. Observe in Figure 8-4a that a poor gyro survey run in a 50 deg hole can have a vertical uncertainty up to ± 7 ft per thousand feet of hole. Therefore, a 10,000-ft well could have a vertical depth error up to ± 70 ft. A 70-ft difference in water level between two wells is significant enough to cause someone to place a production fault, permeability barrier, or water level change on a structure map. The uncertainty charts show that such variations in the depth of a water level in various wells can result from directional survey uncertainties, rather than from geologic problems.

230

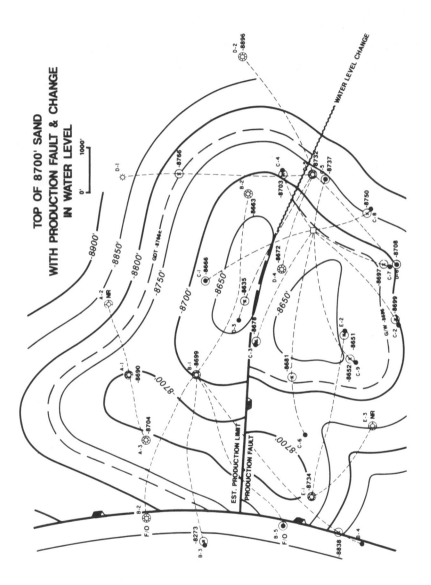

Figure 8-3 Structure map showing arbitrary production fault/water level change to account for differences in gas/water contacts in various wells.

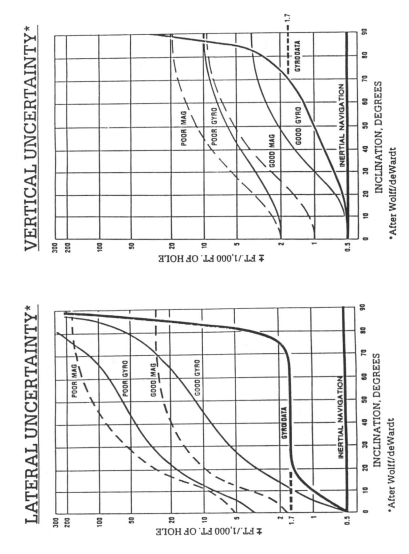

Figure 8-4 (a) Expected vertical uncertainty in a deviated well considering various types of surveys. (b) Expected lateral uncertainty in a deviated well considering various types of surveys. (Modified from Wolff and de Wardt 1981. Published by permission of the Journal of Petroleum Technology and Gyrodata, Inc.)

Checking all geologic possibilities for the water level difference, and verifying that the production fault does not exist, the evidence in this case leaned toward directional well problems. One important piece of evidence was that the water level differences in various wells were not the same, but varied from 20 ft to 80 ft. Figure 8-5 is a reinterpretation of the 8700-ft Sand assuming that the variations in water level are the result of inaccuracies in the directional surveys. Water level differences in the sand vary from $\pm$ 20 ft to $\pm$ 80 ft in 10 wells. Engineering data does not support a production fault or permeability barrier as indicated on the initial interpretation (Fig. 8-3) because the same pressures are recorded in wells throughout the reservoir.

The best way to correct for water level variations in deviated wells is to normalize the water contact to a common water level. If there are any vertical wells in the reservoir that support a common water level, the deviated well data can be corrected to fit the vertical wells. Not only must the water level be corrected, but the top and base of sand will also require correction. These adjustments will alter the configuration of the structure resulting in a more accurate map.

If for some reason, such as time constraints, the water level cannot be normalized, the water contact in each well can be honored in delineating the contact as shown in Figure 8-5. This is a shortcut that does not always result in the most accurate map, but it does provide a more reasonable evaluation of the reservoir than arbitrarily introducing non-existent production faults, permeability barriers, or changes in water level.

In reviewing structure maps and prospects, be aware of the lateral and vertical uncertainties of directional surveys and their effect on a structural interpretation and potential reserves.

DIRECTIONALLY DRILLED WELLS, NET SAND, AND NET PAY ISOCHORE MAPS

True Vertical Thickness (TVT) is that thickness required for the preparation of *Net Sand and Net Pay Isochore Maps*. The true vertical thickness is the thickness of a sand, bed, or formation when measured in a vertical direction or as seen in a vertical well. Directionally drilled wells can have a logged thickness that is thicker than, equal to, or thinner than the thickness in a vertical well drilled through the same section.

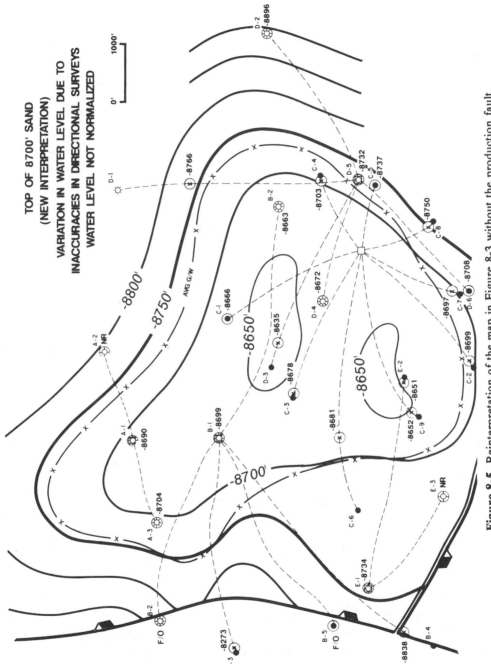

Figure 8-5 Reinterpretation of the map in Figure 8-3 without the production fault.

We have already discussed in an earlier section that a correction factor must be applied to fault cuts when the missing or repeated section is determined by correlation with a directionally drilled well. Likewise, when using thicknesses obtained from directionally drilled wells for net sand or net pay isochore maps, the same correction factor equation must be applied. This equation corrects the exaggerated or directional log thickness to true vertical thickness.

Equation 8-1 shown previously section, is also used for correcting log thicknesses for isochore mapping. Figure 8-6 illustrates the thicknesses seen in a directionally drilled well, compared to the true vertical thickness of the formation. The directional well thicknesses must be corrected to true vertical thickness for net sand and net pay *isochore* maps. Three wells are shown in the figure. Well No. A-1 is vertical, Well No. A-2 is drilled downdip, and Well No. A-3 is drilled in an updip direction. Notice that there is a substantial difference in formation thickness as seen in the measured depth logs when compared to the true vertical depth, true stratigraphic, or true vertical thickness logs.

Based on wrong information or misunderstanding of which bed thicknesses are used for various calculations, a number of geologists, geophysicists, and engineers use the wrong thicknesses for isochore maps and volume calculations. For interval *isopach* mapping and depositional model analysis, the true stratigraphic thickness (TST) for the intervals under study must be used. However, in all cases for reservoir volume calculations, the TVT is required. We have found that some of this misunderstanding comes form work in certain tectonic settings. Refer once again to Figure 8-2. This figure shows that for flat beds the TST or TVDT is the correct thickness to use for volume calculations. This is true only because in flat beds, TST is equal to TVDT, which is also equal to TVT. Say that a geologist has worked in an area where flat beds are common. The geologist will be taught to use TST for vertical wells and a TVDT thickness correction in deviated wells for preparing new sand and net pay isochore maps. This procedure for the flat beds is correct. The geologist is transferred to an area of high dipping beds and begins to work on salt domes or Fault Propagation Folds with high bed dips. There is a tendency for that person to continue to use the techniques he has been taught and used for many years. Therefore, it is not uncommon for the geologist working on an anticline with 40 deg dipping beds to correct deviated well log thicknesses to a TST or TVDT for preparation of net sand and net pay isochore maps. In these cases, the thicknesses are wrong and result in significant volumetric errors.

235

MLT - TVTD - TVT - TST
THICKNESS RELATIONSHIPS

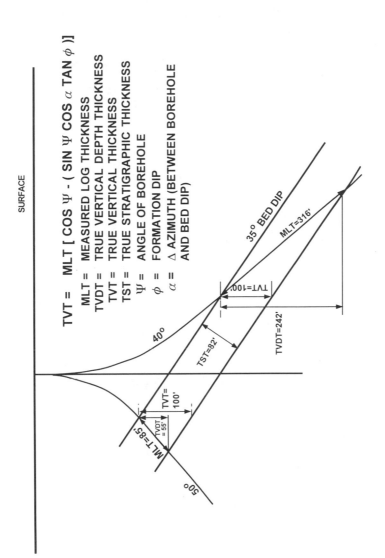

Figure 8-6 Compare the MLT, TVDT, TVT, and TST measurements for the two directionally drilled wells. Consider the implications of using the wrong thickness for volumetric calculations.

Recently, a QLT Seminar was taught to 17 geologic, geophysical, and engineering managers of an oil company. The managers were given a prospect discovery on a highly dipping structure, penetrated by a deviated well. They were asked whether a correction factor was required to calculate the volumetric reserves, and, if so, to what thickness must the measured log be corrected? Or, was the MLT used for the calculations?

All 17 people chose the wrong thickness: 8 chose TST and 9 chose TVDT. Even the reserves manager for the company choose the wrong thickness. He had spent most of his career in the Permian Basin and thought TST was used everywhere to construct net pay isochore maps to calculate reserves.

A recent field study of an offshore Gulf of Mexico field, with many reservoirs, revealed that three different thicknesses (TST, TVDT, TVT) were used in the same field to prepare the isochore maps and calculate individual reservoir volumes. The thickness used for a specific reservoir depended upon the person who did the work. And, the choice of thickness was based on what each individual had been taught and where they had worked prior to their current work in the United States Gulf of Mexico.

In our experience, one of the largest volumetric errors we have seen resulted from the use of MLT from a deviated well to prepare a net pay isochore map for a dipping reservoir. The MLT used was 750 ft of gas, which resulted in a very large gas volume. When it was later recognized that the wrong thickness was used and the TVT calculated, the thickness reduced to 110 ft. This change significantly reduced the reserves assigned to this reservoir. Such mistakes are very costly and must be avoided.

CHAPTER NINE

ISOCHORE MAP
QUICK LOOK TECHNIQUES (QLTs)

INTRODUCTION

For any given prospect or producing reservoir, estimates of the hydrocarbons in place are obtained from net pay isochore maps. These maps delineate the true vertical thickness of reservoir quality rock which is required to determine the volume of hydrocarbons in terms of acre-feet.

These isochore maps must be constructed accurately in order to represent true hydrocarbon volume. In this chapter, we present QLTs related to evaluating completed isochore maps.

THE WHARTON METHOD FOR NET PAY ISOCHORE MAPS

An incorrectly constructed net pay isochore map can result in an unrealistic overestimate or underestimate of hydrocarbon reserves. One of the most common mapping errors involves the construction of *Edge Water Isochore Maps* (Tearpock and Bischke 1991). The construction of an isochore map for an edge water reservoir is more complex than the construction of a bottom water isochore. J. B. Wharton (1948) developed a very accurate method for constructing these isochore maps that is still used today.

The primary error in construction of edge water reservoir, net pay isochore maps, involves the contouring *method used to connect* the isochore contours within the full thickness area to those in the water wedge zone. The method is critical and requires special attention. To prepare an edge water net pay isochore map, the data needed are:

1. structure map - top of porosity;
2. structure map - base of porosity;
3. net sand isochore map;
4. net pay values for all wells; and
5. the subsea depth of fluid contacts: hydrocarbon/water contact on the top and base of sand; gas/oil contact.

The fluid contact on the base of sand, projected onto the top of the structure map, is called the (ILW) Inner Limit of Water (Tearpock and Harris 1987; Tearpock and Bischke 1991).

A net pay isochore map for an edge water reservoir (single hydrocarbon phase reservoir - oil or gas) is divided into two areas: *the full thickness area, and the water wedge zone* (Fig. 9-1). These two areas are separated by the ILW (hydrocarbon/water contact on the base of sand). The entire isochore is bounded by the zero line which can either be the hydrocarbon/water contact on the top of the sand, a fault, shale out, salt or permeability barrier. Within the full thickness area, net pay is equal to net sand, since the formation is full from top to base with hydrocarbons (Fig. 9-1a). The wedge zone contains both hydrocarbons and water. Unlike the full thickness area which is controlled by net sand contours (Fig. 9-1b), the major influences for contouring the wedge are the structural attitude of the formation and the shale distribution within the formation.

The inner limit of water is the updip limit of water within the reservoir. Therefore, within the full thickness area (area updip of the ILW - full from top to base with hydrocarbons), the net pay contours are the same as the net sand contours. Some or all of these full thickness or net hydrocarbon contours, which equal the net sand contours, intersect with the inner limit of water. *The full thickness contours cannot continue unaffected past the inner limit of water into the wedge zone because downdip from the ILW the sand contains both hydrocarbons and water.* We know, however, that the isochore contours, within the full thickness area which intersect the ILW, must close. Therefore, they must enter the wedge zone. The question is how and where? Figure 9-2 shows a cross section parallel to the 50-ft net sand contour shown in Figures 9-1a and 9-1b. Along the 50-ft net sand contour line, the amount of reservoir quality sand is 50 ft. Above the gas/water contact on the base of sand, net sand is equal to net gas. None of the sand is wet. Between the gas/water contact on the base of the sand and the gas/water contact on the top of sand in a downdip direction, the 50-ft of net sand contains a decreasing amount of gas and an increasing amount of water. Therefore, this 50-ft net gas contour cannot continue along the 50-ft net sand contour line from the full thickness area, through the ILW, downdip into the water wedge. We know, however, that the 50-ft contour must enter and continue in the wedge zone, but where? If the 50-ft contour is to continue into the wedge, it must do so where there is more net sand than 50 ft. Since some of the reservoir sand in the wedge is wet, the 50-ft net gas contour must exist where there is more than 50 ft of net sand. *Therefore, the strike direction of the contour must abruptly turn, at the inner limit of water, toward the thicker sand or next higher net sand contour (60-ft).* Figure 9-1a illustrates how the 50-ft contour abruptly turns at the ILW.

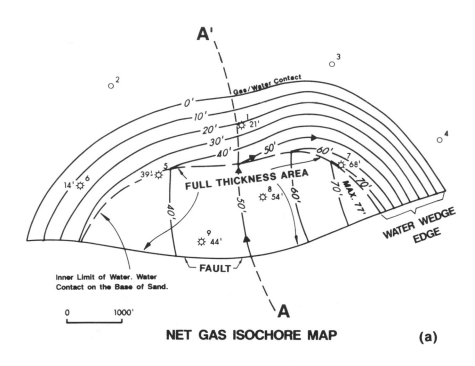

NET GAS ISOCHORE MAP   **(a)**

Figure 9-1a Net gas isochore map of an edge water reservoir constructed correctly using the Wharton Method.

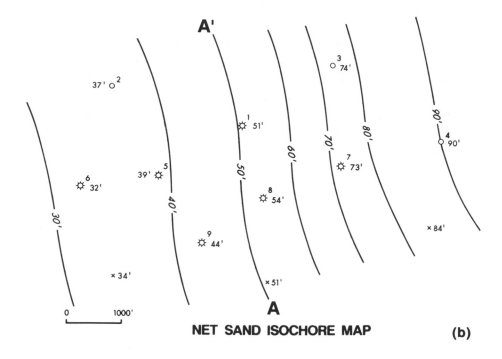

A

NET SAND ISOCHORE MAP

(b)

Figure 9-1b Net sand isochore map used to construct the net pay map shown in Figure 9-1a.

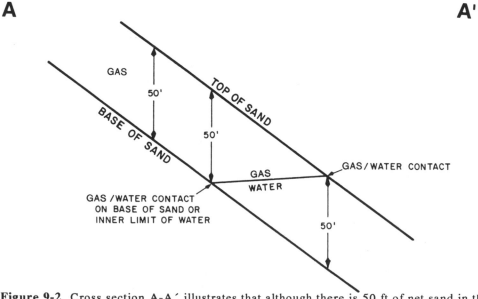

Figure 9-2 Cross section A-A´ illustrates that although there is 50 ft of net sand in the reservoir downdip of the ILW, some of the sand is wet. Therefore, 50 ft of net gas cannot exist along this cross section in the wedge zone.

If the 50-ft contour or any other contour within the full thickness area is continued, unaffected downdip of the inner limit of water, the volume of hydrocarbons estimated for the reservoir will be overestimated. An example of this type error is shown in Figure 9-3a. From the full thickness area, updip of the inner limit of water, the net pay contours continue, without change, into the wedge zone. This error results in an incorrect map which overestimates the hydrocarbon volume. Figure 9-3b employs the correct use of the Wharton Method, where the full thickness contours do not continue unaffected in the wedge, but abruptly change their strike direction at the inner limit of water toward the thicker sand.

The mapping error discussed in this section is a common error made in the preparation of net hydrocarbon isochore maps. The volume planimetered for Figure 9-3a is 13,937 ac-ft, in contrast to 12,401 ac-ft for the correctly contoured map (Fig. 9-3b). This results in a 1,536 ac-ft overestimate of reservoir volume. Considering 500 barrels of oil per ac-ft recovery factor, the overestimate is equal to 768,000 barrels of oil.

How can you check for this type of error? If the isochore map is prepared as shown in Figure 9-3a, with the inner limit of water shown on the map, it is relatively easy to check to see whether or not the full thickness contours have taken an abrupt turn at the ILW. If the dashed ILW has been *erased*, as it is on many maps, it will be necessary to underlay the structure map on the base of sand, over the net hydrocarbon isochore map and trace in the ILW. Next, observe if the net pay isochore contours turn at the ILW. If they do not, then the isochore map should be remapped before being planimetered to determine the correct hydrocarbon volume.

TOO MUCH NET PAY

It is not uncommon to see a net pay isochore map with contours reflecting more pay then can be present in the reservoir. This error can occur by mapping more pay than actual net sand present or by mapping more pay than the amount of net sand above a water contact. Such errors usually result from carelessness rather than from a misunderstanding of proper methods. Nevertheless, the errors can result in incorrect estimates of potential or proved reserves.

242

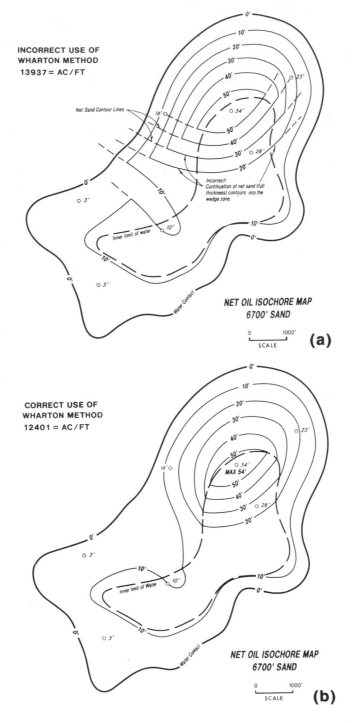

Figure 9-3 (a) Net oil isochore map constructed incorrectly resulting in an overestimate of reserves. (b) A correctly constructed net oil isochore map using the same data as in 9-3a and the Wharton Method.

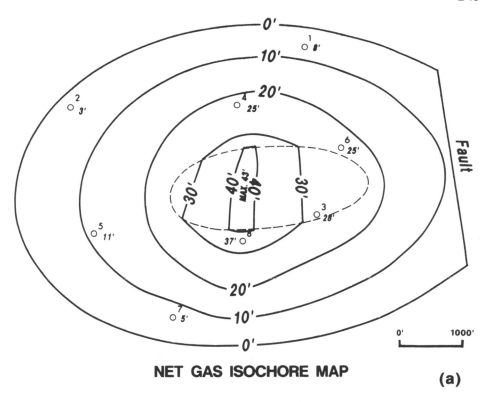

NET GAS ISOCHORE MAP

(a)

Figure 9-4a Net gas isochore map of an edge water reservoir.

Figures 9-4a and 9-4b each show a completed net gas isochore map for a reservoir; however, they are slightly different. Can you tell what the difference is between the two net pay maps?

Look at the 10-ft and 20-ft contour lines. In the western portion of the reservoir in Figure 9-4b, the two contours extend further to the west, closer to the zero line, when compared to the map in Figure 9-4a, resulting in a larger reservoir volume. In the eastern portion of the reservoir, the 20-ft contour makes a peculiar abrupt change in strike direction in Figure 9-4b and is placed further to the east than the 20-ft contour in Figure 9-4a.

Can you now see a problem with the net pay isochore map in Figure 9-4b? If this were a reservoir that you were evaluating, how could you tell if the isochore map was constructed correctly, and if the assigned pay corresponds to the net sand in this area? The answer is to overlay the net pay isochore map onto the net sand map (Fig. 9-5) and evaluate the net pay contours.

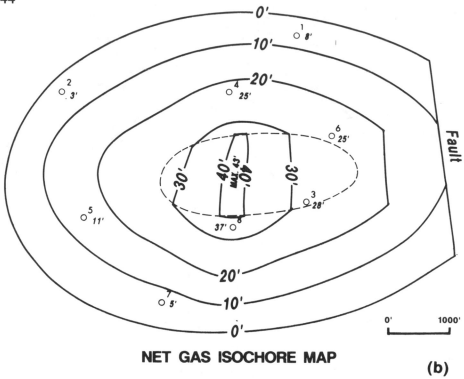

NET GAS ISOCHORE MAP

(b)

Figure 9-4b Same reservoir shown in Figure 9-4a. The construction is slightly different.

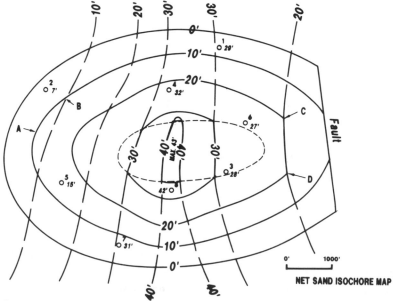

NET SAND ISOCHORE MAP

Figure 9-5 Net gas isochore map shown in Figure 9-4b, with the net sand isochore map as an overlay. Review Points A, B, C, and D.

Two different errors were made in the preparation of the net gas isochore map shown in Figure 9-4b. Let's first look at the 10-ft and 20-ft net gas and net sand contours in the western portion of the reservoir (Fig. 9-5). Notice that the 10-ft and 20-ft net gas contours cross the net sand contours of the same value. The 10-ft and 20-ft net gas contours are in the water wedge, downdip of the inner limit of water. Therefore, the net sand in this position of the reservoir contains both gas and water. At Point A, however, the map indicates that there is 10 ft of gas pay even though the sand is less than 10 ft thick ($\pm$ 7 ft) and a portion of the sand is gas filled, and a portion contains water. **This is impossible.** You can't have more pay than net sand. The same error is also made with the 20-ft net sand and net gas contours. *This error results in an overestimate of the gas volume.*

The second error occurs in the eastern portion of the reservoir. We mentioned earlier that the 20-ft net gas contour makes a peculiar, abrupt change in strike direction. If we again look at Figure 9-5, we see that the 20-ft net gas contour intersects the 20-ft net sand contour (Point C) and then parallels the net sand contour to Point D, where it again changes strike direction.

What is the contouring problem in the area from Points C to D? The net gas contour intersects and follows the net sand contour line *downdip* of the inner limit of water, within the water wedge zone. This cannot happen; it is incorrect. Remember, only *updip* of the inner limit of water, where the net sand is completely filled with pay, can a net pay contour line follow or equal a net sand contour line. Within the water wedge (Fig. 9-5) one cannot contour 20 ft of gas as shown on the map, since the 20 ft of net sand contains both water and gas. In this case, the amount of sand was not overestimated; instead, the amount of sand containing pay above the water contact was overestimated.

To quickly check net pay isochore maps for accuracy, a net pay isochore map under review must be overlaid onto a net sand isochore map, as shown in Figure 9-5. In this manner, incorrect contouring on the net pay map can be identified. If a net sand map was not prepared, it may be necessary to construct a net sand map in order to check the accuracy of the net pay isochore maps. An overestimated reservoir volume will have a negative impact on project economics.

ISOCHORE VERSUS ISOPACH

Ever since the boom days in our industry, there has been widespread use of the word "isopach" to describe thickness maps whether the maps portray vertical or stratigraphic thickness data. This usage is not technically correct. An isochore map delineates the **true vertical thickness** of a rock unit, while an isopach map illustrates the **true stratigraphic thickness** of a rock unit. This means that an isopach map is different from an isochore map.

Problems related to volumetric calculations of hydrocarbon reserves commonly occur because of the misunderstanding related to these two terms. Net sand and net pay maps are **always** prepared using true vertical thickness values from well log data. Therefore, net sand and net pay maps are isochores rather than isopachs. Because of incorrect and limited training during the boom days, many people believe that volumetric calculations, for oil and gas, are based on isopach, rather than isochore values. This leads to incorrect volume calculations.

The title blocks of nearly all net sand and net pay maps that are prepared by geoscientists and engineers are labeled as isopach maps. Does this title mean that the mapper actually used stratigraphic thickness instead of vertical thickness for the map or is the use of the word isopach just an incorrect grammerical substitution for isochore? The only way to answer the question is to ask the mapper several questions. What do you mean by isopach? What thickness did you use for the net pay map? The answers to these questions can be critical with respect to oil and gas volumes.

All net sand and net pay maps must use true vertical thickness values and therefore, should be titled "isochores". We recommend the use of the word isochore on all title blocks for net sand and net pay maps. This usage should eliminate the inconsistencies and current problems encountered with the use of the words isochore and isopach.

Figure 9-6 illustrates the difference between an isopach map and an isochore map. The figure from Bucher and Hintze, 1962, shows that the words isochore and isopach were correctly understood years ago. When did the understanding and usage of these terms change? We believe it occurred during the boom-time and has plagued the industry ever since. Review the figure carefully. The correct understanding of these terms can be critical to your reserve estimates and economics.

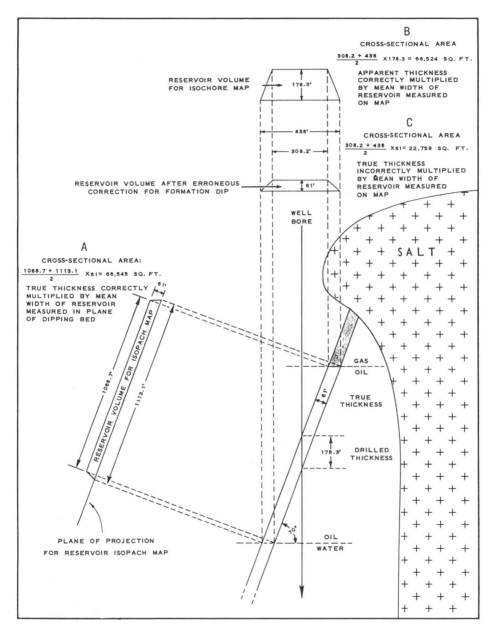

Figure 9-6 Computation of reservoir volume from an isopach map projection (A) and from an isochore map projection (B). Compare these with Computation C, where the formation thickness has been corrected erroneously from the drilled thickness to the true thickness. (From Bucher and Hintze, 1962. Published by permission of Exxon USA Company)

CHAPTER TEN

PROJECT EVALUATIONS

INTRODUCTION

Throughout the textbook we have presented numerous QLTs that can be used to review and evaluate a prospect before drilling. QLTs can often provide accurate and rapid answers about the quality of a prospect or geologic interpretation. These answers can be invaluable when deciding whether or not to invest in a prospect.

In this chapter, we present a number of completed prospect maps. Your job is to apply whatever QLTs you determine are applicable to evaluate the quality of the completed work and prospects. QLTs are sometimes just important questions to be answered; for example: were all the data used, were the required maps prepared, or were the proper thicknesses used for isochore mapping.

QUICK LOOK TECHNIQUE PROSPECT REVIEW QUESTIONS

When reviewing geologic interpretation or prospects, some of the QLTs used to conduct the evaluation may be as simple as asking the right questions. We recommend that each evaluator develop a list of questions to ask to the generator or team. These questions can often speed up the review process and pinpoint specific areas that require detailed review.

A sampling of QLT questions follows. The questions are divided into specific topics. An evaluator can customize this list for each situation.

QLT QUESTIONS

I. Data

A.   Were all the data used?

      1.   Electric logs, seismic sections, production, dipmeters, etc.

B.   What data were not available?

II. Contouring

A.   What contouring method was used to make the maps?

B.   Are the maps contoured optimistically or pessimistically?

C.   Are the prospect highs "tee pee" structures?

D.   Do all contours close (odd number of contours around a finite fault)?

E.   Is there contour compatibility across faults?

III. Faulting

A.   Are fault surface maps constructed?

B.   Are the fault surface maps geologically and geometrically reasonable?

C.   Do the faults conform to those expected in the tectonic setting?

D.   Are the faults syndepositional?

E.   Was growth analysis done on the faults?

F.   Are the faults downward dying?

IV. Structure Mapping

A.   How many horizons were mapped?

B.   Were fault maps integrated (cross contoured) with the structural horizons?

C.   Was the vertical separation used to contour across faults?

D.   Is there structural compatibility between closely spaced horizons?

E.   Is there structural compatibility across small faults?

F.   Were restored tops used in preparing the structure maps?

G.   Are porosity top maps required? Were they made?

H.   Were all relevant surfaces mapped?

I.   Is fault displacement conserved at fault intersections?

V. Isochore Mapping

   A.   Is the reservoir a bottom water or edge water reservoir?

   B.   Edge Water Reservoirs:

       1.   Was the Wharton Method used?

       2.   Was the wedge constructed correctly?

       3.   Are the full thickness contours connected correctly to the wedge edge contours?

       4.   Is the inner limit of water shown on the map?

       5.   Was the net sand isochore map used to guide the construction of the wedge edge net pay contours?

   C.   Was TVT used for the net sand and net pay maps?

   D.   What correction factor was used to convert deviated well log thicknesses?

   E.   Is the net pay isochore based on a top of structure or top of porosity map?

VI. Structural Geology

   A.   Extensional:

       1.   Does fault shape relate to fold shape?

       2.   Does the normal fault turn down (antilistric)?

       3.   Are the faults downward dying?

       4.   Is the fault real or is it an axial surface?

       5.   Does the prospect high tie to a flat on the fault surface?

       6.   On seismic data are faults drawn through coherent reflectors?

   B.   Compressional:

       1.   Are the faults and folds geometrically related?

       2.   Do formations maintain unit thickness?

       3.   Are cross sections balanced?

       4.   Are the thrust faults drawn with a continuously listric shape?

       5.   Are faults drawn through no data zones?

       6.   What's the structural style of the area?

       7.   What types of folds are present in the area?

VII. Seismic Data

       1.   Has all the seismic data been used?

       2.   Are there any statics problems in the area?

3. Are wells tied to the seismic lines?

4. How were wells projected into the seismic lines?

5. Is there a good time-depth velocity function for the area?

6. Have faults been pushed through coherent reflectors?

7. Are any amplitudes present on the seismic lines?

VII. General

A. Were the log and seismic correlations loop tied?

B. Were all the seismic lines used?

C. Were cross sections made?

PROJECTS

You are in a budget presentation to review various prospects, maps, and fields. You must make rapid judgmental decisions on which properties to buy, which prospects you should drill, which maps are correct or incorrect, and which maps have reasonable interpretations. Each project presented is different; however, all the information or data required to make a decision is available on the maps or in the writeup on each project. Your task is to evaluate each project and make a sound decision whether or not to participate in the project.

Each project is described, and relevant data are included. In the space provided below each project description, list the QLTs that you decide to apply. If the QLT is a simple question, write it in the space available. Test the interpretation by applying the QLTs. Finally, make your decision on the project.

Good Luck!

PROJECT ONE

Prospects are in a normal growth fault environment. Two wells are proposed to drain two untested fault blocks in between and fault separated from two producing fault blocks. Probability of success is 75%. Both well log and seismic data were used to generate the interpretation.

Question: Based on the structure map, is it geologically reasonable to approve either location? Yes or no? Why?

List QLT questions:

Identify mapping problems, if any:

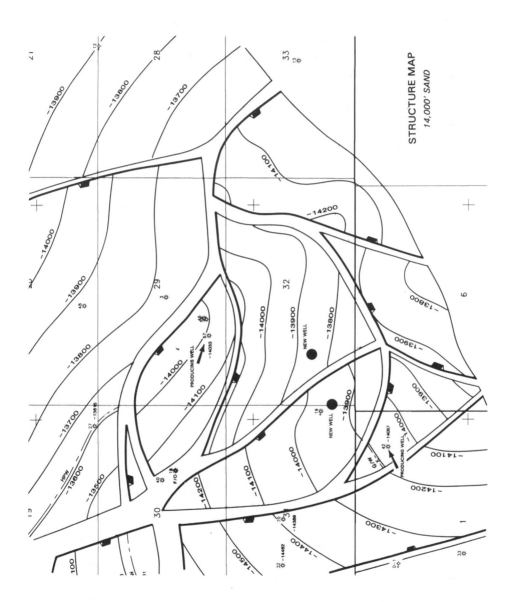

STRUCTURE MAP
14,000' SAND

PROJECT TWO

The prospect is on the Eastern flank of a salt dome. Well No. 2 is producing oil from this zone. A well is proposed to be drilled updip of Well No. 2 and upthrown to Fault E-1. Downthrown to Fault E-1 Well No. 7 is wet with a HKW at -9447 ft.

Question: Considering the fault and structure data provided would you approve the location without additional work? Yes or no? Why?

List QLT questions:

Identify mapping problems, if any:

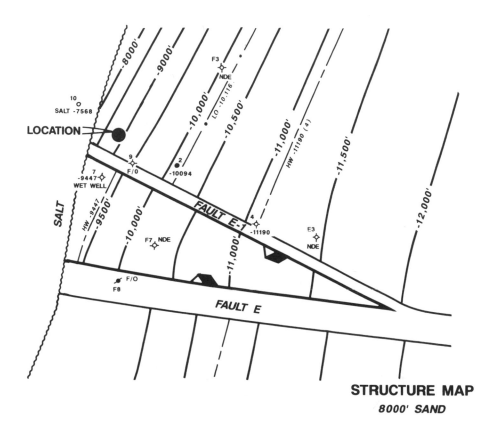

STRUCTURE MAP
8000' SAND

PROJECT THREE

The structure map shows a prospect in the hanging wall of a thrust fault. Observe that there is a producing, hanging wall structure to the north of the prospect block. The prospect is based on seismic data. The estimated oil/water contact is at 15,500 ft.

Question: Evaluate the map to determine if you would like to participate in the prospect.

List QLT questions:

Identify mapping problems, if any:

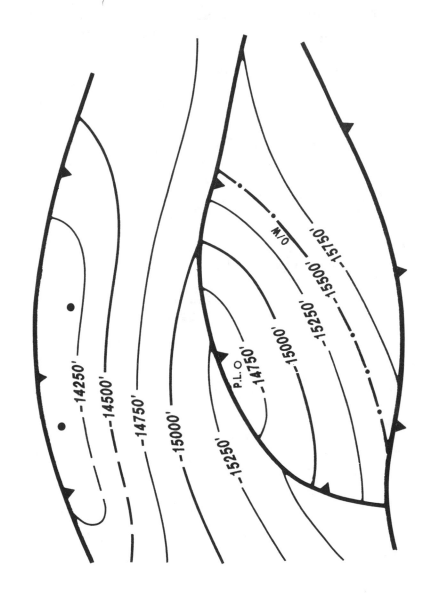

THRUST PROSPECT

PROJECT FOUR

You are drilling a critical well. The required pay for an economic reservoir is 135 ft. The well is being drilled directionally into a dipping structure. The data for the well and structure are provided below.

Question: Is the well an economic success, or do you sell the discovery?

Consider if a correction factor is required to analyze the net pay. Yes or no? If no, what is the value for net pay? If yes, which thickness is used for net pay values (MLT, TVDT, TVT or TST) and which equation must be used to determine the correct pay thickness?

Hint: Draw a cross section.

Data:

Wellbore Angle	= 35°
Borehole Compass Direction	= 80° E
Formation Dip	= 42°
Formation Dip Compass Direction	= N 80° W
Measured Log Thickness of Pay	= 142 ft

PROJECT FIVE

Well No. 47 has a show. Plans are to drill a well updip, as shown on the map. The prospect is trapped upthrown to two faults.

Question: Evaluate the map and determine if the prospect is reasonable.

List QLT questions:

Identify mapping problems, if any:

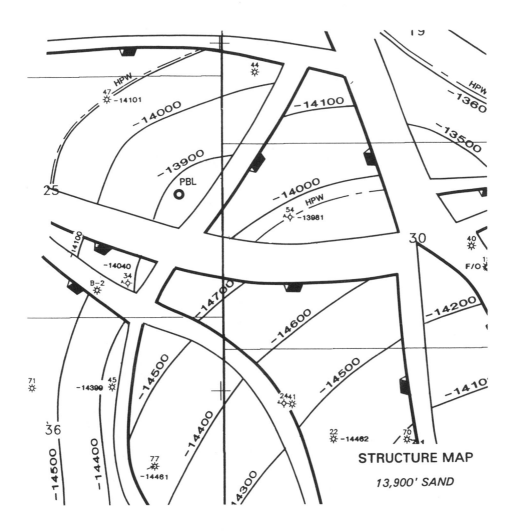

STRUCTURE MAP

13,900' SAND

PROJECT MAPS SIX THROUGH TEN

Identify as many mapping mistakes as you can using Quick Look Techniques.

PROJECT MAP SIX

Identify mapping mistakes, if any:

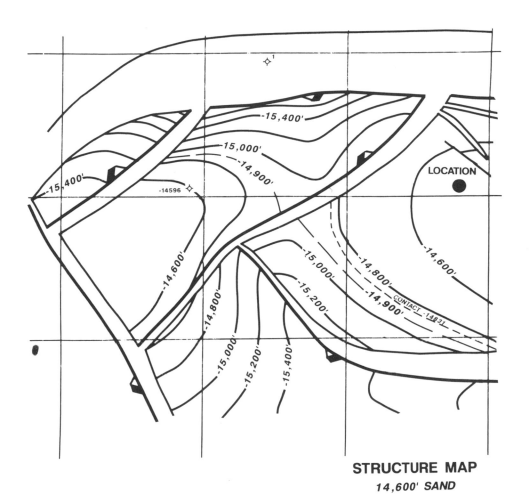

STRUCTURE MAP
14,600' SAND

PROJECT MAP SEVEN

Identify mapping mistakes, if any:

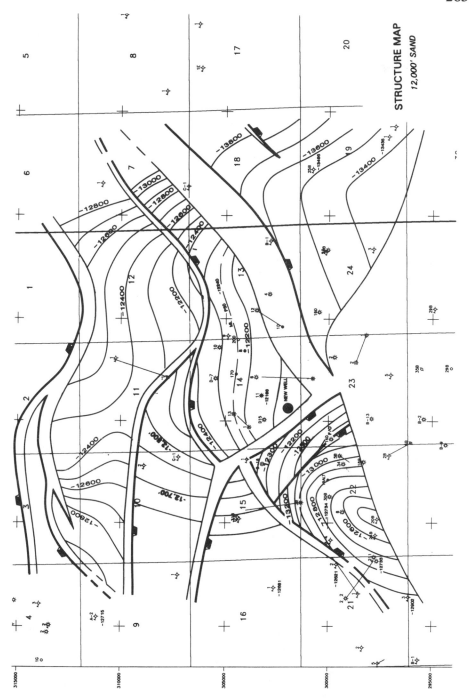

STRUCTURE MAP
12,000' SAND

PROJECT MAP EIGHT

Identify mapping mistakes, if any:

PROJECT MAP NINE

Identify mapping mistakes, if any:

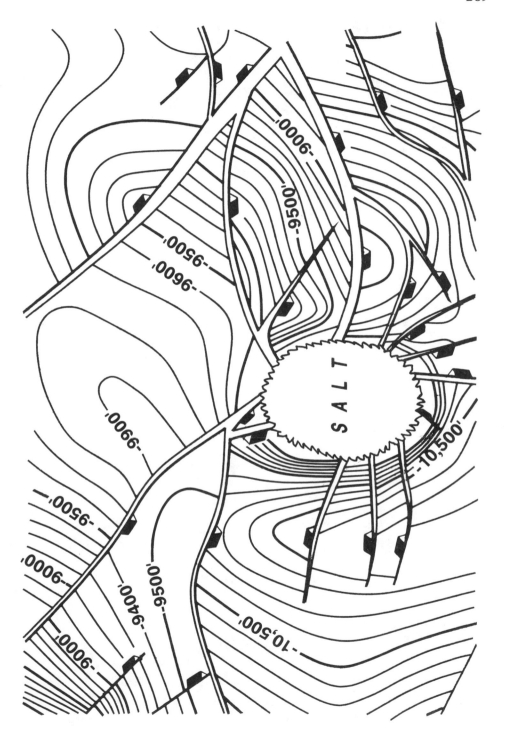

PROJECT MAP TEN

Identify mapping mistakes, if any:

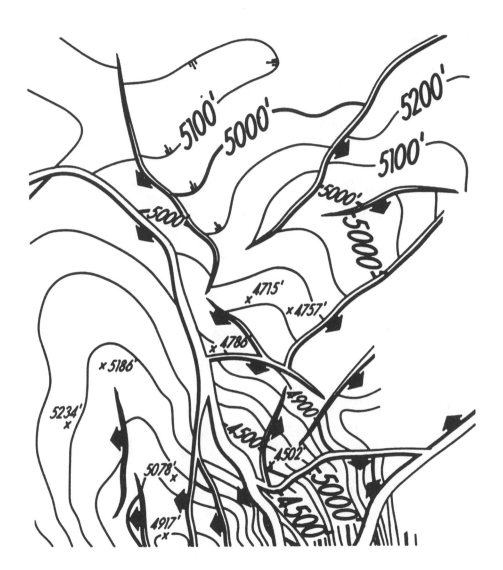

PROJECT MAP ELEVEN

Examine the structure map. Plans are to drill development wells updip of the producing oil wells. The company wants to shoot one or more additional seismic lines to obtain more information about Fault A.

Questions: 1. How would you layout the seismic lines?
2. In what direction is the fault striking?
3. Can you generate an implied fault surface map from this structure map.
4. Explain what is going on three-dimensionally?

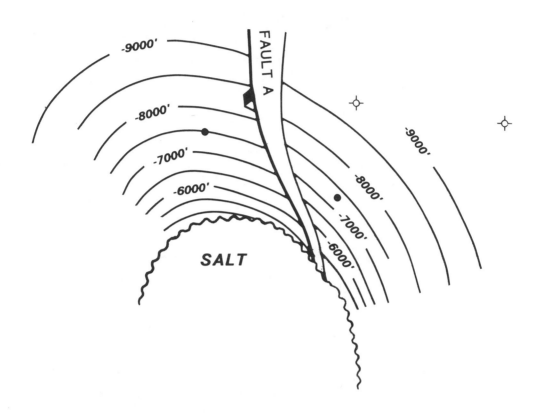

STRUCTURE MAP
6000' SAND

0'   1000'

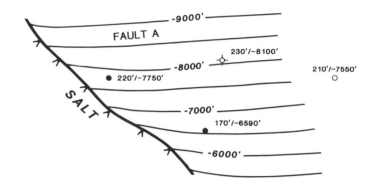

FAULT SURFACE MAP

FAULT A

0'   1000'

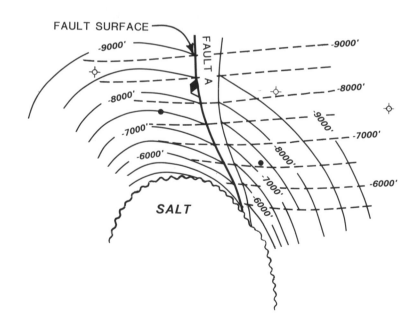

STRUCTURE MAP

6000' SAND

0'   1000'

PROJECT MAP TWELVE

Review the structure map. Explain why the gaps for the two faults are different.

Questions: 1. Define the relationship, if any, between the two faults.
2. When will the two faults intersect?
3. Can you construct an implied fault surface map?

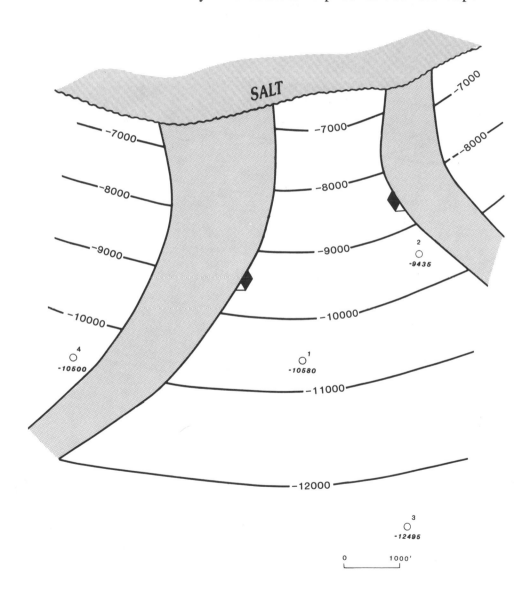

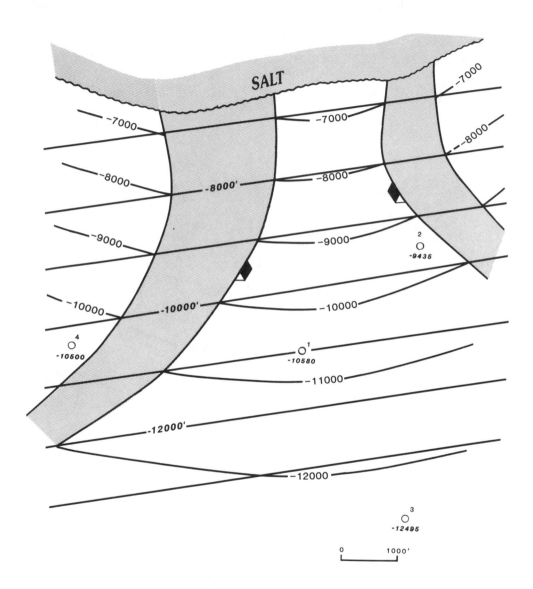

REFERENCES

Allan, U. S., 1989, *Model for hydrocarbon migration and entrapment within faulted structures,* AAPG Bull., v. 73, No. 7, p. 803-811.

Atwater, G. I. and E. E. Miller, 1965, *The effect of decrease in porosity with depth on future development of oil and gas reserves in South Louisiana,* AAPG Bull., v. 49, p. 334.

Bally, A. W., 1983, *Seismic Expression of Structural Styles,* AAPG Studies in Geology, Series No. 15, v. 3.

Banks, C. J., and J. Warburton, 1986, *"Passive-roof" duplex geometry in the frontal structures of the Kirthar and Sulaiman Mountain Belts:* Jour. Structural Geol., v. 8, p. 229-237.

Bell, W. G., 1956, *Tectonic Setting of Happy Springs and Nearby Structures in the Sweetwater Uplift Area,* Central Wyoming: in Geological Record of the AAPG, Rocky Mountain Section.

Bischke, R. E., 1990, *Applied structural balancing:* GCAGS--Short Course (GCAGS 40th Annual Convention in Lafayette, LA).

---, 1992, *Applied subsurface mapping,* short course notes, Subsurface Consultants & Associates, Inc., Lafayette, LA.

---, 1994, *The compressional off structure problem,* HGS Bull., May, 1994

---, 1994, *Interpretation of sedimentary growth structures from well log and seismic data,* AAPG Bull., v. 78, p. 873-892.

Bischke, R. E., and S. F. Elston, 1991, *Quantitative analysis of sedimentary growth structures,* GSA Abstracts, p. A421.

Bischke, R. E., and J. Suppe, 1990, *Calculating sandstone/shale ratios from growth normal fault dips on seismic profiles:* Trans. - GCAGS, v. 40.

---, 1990, *Geometry of rollover-origin of complex arrays of small antithetic and synthetic faults,* AAPG Bull., v. 74, p. 611.

---, 1993, *Origin of complex arrays of crestal growth faults,* in review, to be submitted to AAPG Bull.

Bischke, R. E., and D. J. Tearpock, 1993, *A method for estimating gross sand percentage and reservoir thicknesses from seismic sections: example from Segno Field, Polk, Texas,* HGS Bull., May, 1993, p. 22.

Bishop, M. S., 1960, *Subsurface Mapping,* John Wiley & Sons, New York, NY.

Boyer, S. E., 1986, *Styles of folding within thrust sheets: examples from Appalachian and Rocky Mountains of the U.S.A. and Canada:* Jour. of Struct. Geol., v. 8, p. 325-339, Pergamon Press Ltd.

Boyer, S. E., and D. Elliott, 1982, *Thrust systems:* AAPG Bull., v. 66, No. 9, p. 1196-1230.

Bucher, W. H., and W. H. Hintze, 1962, *Contouring Techniques for Structure-Contour and Isopach Maps,* Figure 66.

Coffeen, J. A. 1984, *Interpreting Seismic Data,* PennWell Publishing Co., Tulsa, OK.

Crowell, J. C., 1974, *Origin of late Cenozoic Basins in Southern California, in Tectonics and Sedimentation:* Society of Econ. Paleo. and Min., Special Pub. No. 22, p. 190-204.

Curtis, D. M., 1984, *Finding deep sands in the Gulf Coast Tertiary,* Houston Geol. Soc., Continuing education committee course notes, p. 1-72.

Dalstrom, C.D.A., 1969, *Balanced cross sections:* Can. Jour. Earth Sci., v. 6, p. 743-757.

Dickinson, G., 1953, *Geological aspects of abnormal reservoir pressure in Gulf Coast Louisiana,* GSA Bull., v. 37, p. 410-432.

Dula, W. F. JR., 1991, *Geometric models of listric normal fault rollover folds,* AAPG Bull., v. 75, p. 1609-1625.

Elliott, D., 1976, *The energy balance and deformation mechanisms of thrust sheets:* Phil. Trans. Roy. Soc. London, Part A, v. 283, p. 289-312.

Fisher, W. L. and J. H. McGowen, 1967, *Depositional systems in the Wilcox Group of Texas and their relationship to occurrence of oil and gas,* Trans. - GCAGS, v. 27, p. 105-125.

Gibbs, A.D., 1983, *Balanced cross section construction from seismic sections in areas of extensional tectonics:* Jour. Structural Geol., v. 5, No. 2, p. 153-160.

Goguel, J., 1962, *Tectonics,* second ed., W.H. Freeman and Co., San Francisco, CA.

Hamblin, W. K., 1965, *Origin of "reverse drag" on the downthrown side of normal faults:* GSA Bull., v. 76, p. 1145-1164.

Hamilton and Jones, 1992, *Computer modeling of geologic surfaces and volume,* AAPG Computer Applications in Geology, No. 1.

Harding, T. P., 1990, *Identification of wrench faults using subsurface structural data: Criteria and pitfalls,* AAPG Bull., v. 74 No. 10, p. 1590-1609.

Hubbert, M. K., and W. W. Rubey, 1959, *Role of fluid pressure in mechanics of overthrust faulting:* GSA Bull., v. 70, p. 115-166.

Jaeger, J. C., 1962, *Elasticity, fracture and flow with engineering and geological application,* Methuen & Co. Ltd, London, p. 268.

Jones, P. B., 1971, *Folded faults and sequence of thrusting in Alberta Foothills:* AAPG Bull., v. 55, p. 292-306.

---, 1982, *Oil and gas beneath east-dipping underthrust faults in the Alberta Foothills,* Rocky Mtn. Assoc. of Geol., v. 1, p. 61-74.

Lamerson, P.R., 1982, *The fossil basin and its relationship to the Absaroka thrust system,* Wyoming and Utah, in R.B. Powers, ed., Geologic studies of the Cordilleran Thrust Belt: Denver, Rocky Mountain Association of Geologist.

Laubscher, H.P., 1977, *Fold Development in the Jura:* Tectonophy, v. 37, p. 337-362.

Leach, W. G., 1993, *Fluid migration, HC concentration in South Louisiana tertiary sands,* Oil and Gas Journal, May 15, 1993, p. 71.

Lowell, J. D., 1985, *Structural Styles in Petroleum Exploration,* OGCI Publication, Tulsa, OK.

Marshak, S., and G. Mitra, 1988, *Basic Methods of Structural Geology,* Prentice Hall, Englewood Cliffs, NJ.

Mitra, S., 1986, *Duplex structuures and imbricate thrust systems: geometry, structural position, and hydrocarbon potential:* AAPG Bull., v. 70, p. 1087-1112.

---, 1988, *Three-dimensional geometry and kinematic evolution of the Pine Mountain thrust system southern Appalachians:* GSA Bull., v. 100, p. 72-95.

---, 1990, *Fault propagation folds: geometry, kinematic evolution, and hydrocarbon traps,* AAPG Bull., v. 74, p. 921-945.

Mitra, S., and Q. T. Islam, 1991, *Experimental models of the geometry and kinematics of inversion structures,* GSA Abstracts, p. A105.

Morley, C.K., and et. al., 1990, *Transfer zones in the east African rift system and their relevance to hydrocarbon exploration in rifts:* AAPG Bull., v. 74, No. 8, p. 1234-1253.

Namson, J., 1981, *Structure of the western Foothills Belt, Miaoli-Hsinchu area Taiwan: (1) Southern Part:* Petroleum Geol. Taiwan, No. 18, p. 31-51.

Narr, W. and J. Suppe, 1993, *Kinematics of basement involved compressional structures,* Am. Jour. of Science, in press.

Nunns, A. G. 1991, *Structural restoration of seismic and geologic sections in extensional regimes,* AAPG Bull., v. 75, p. 278-297.

Ocamb, R. D., 1961, *Growth faults of south Louisiana:* Trans. - GCAGS, v. 11, p. 139-175.

Pacht, J. A., B. Bowen, B. L. Shaffer, and W. R. Pottorf, 1992, *Systems tracts, seismic facies, and attribute analysis within a sequence - stratigraphic framework; example from offshore Louisiana Gulf Coast, in marine clastic reservoirs,* e.g. Rhodes and T. F. Moslow, Springer Verlag, 1992, p. 21.

Palmer, H.S., 1919, *New graphic method for determining the depth and thickness of strata and the projection of dip: in Shorter Contributions to Geology* - 1918, U.S. Geol. Survey Prof. Pap. 120, p. 122-128.

Price, L. C., 1980, *Utilization and documentation of vertical oil migration in deep basins,* Jour. Petrol. Geol., v. 2, p. 353-387.

Rainwater, E. H., 1963, *The environmental control of oil and gas in terrigenous clastic rocks,* Trans. - GCAGS, v. 13, p. 79-94.

Ramsey, J. G., 1967, *Folding and Fracturing of Rocks,* McGraw-Hill Inc., New York, NY.

Rich, J. L., 1934, *Mechanics of low-angle overthrust faulting as illustrated by Cumberland thrust block, Virginia, Kentucky, and Tennessee:* AAPG Bull., v. 18, p. 1584-1596.

---, 1951, *Three critical environments of deposition and criteria for recognition of rocks deposited in each of them,* GSA Bull., v. 62, p. 1-20.

Roux, W. F., Jr., 1978, *The development of growth fault structures:* AAPG, Structural Geology School Notes.

Setchell, J., 1958, *A nomogram for determining true stratum thickness:* Shell Trinidad, EP 28884, Abstract in PA Bull., No. 127/128, N.V. DeBataafache Petroleum Maatschappij, The Hague, Production Dept., p. 8.

Shaw, J., R. E. Bischke, and J. Suppe, 1994, *Relations between folding and faulting in the Loma Prieta Epicentral Zone: Strike-slip fault-bend folding,* The Loma Prieta, Califormia, Earthquake of October 17, 1989, Tectonic Processes and Models, U.S. Geol. Survey Prof. Paper 1550-F, p. F3-F21.

Smith, D. A., *Sealing and nonsealing faults in LA Gulf Coast Salt Basin*, February, 1980, v. 64, No. 2, p. 145-172.

Spand, J. H., J. P. Evans, and R. R. Berg, 1985, *Balanced cross sections of small fold-thrust structures*: Mountain Geologist, v. 22, p. 41-46.

Stone, D. S., 1991, *Identification of wrench faults using subsurface structural data: Criteria and pitfalls: Discussion*: AAPG Bull., v. 75, p. 1784-1785.

---, 1991, *Analysis of Scale Exaggeration on Seismic Profiles,* AAPG Bull. No. 7, v. 75, p. 1161-1177.

Suppe, J., 1983, *Geometry and kinematic of fault-bend folding:* Am. Jour. Sci., v. 283, p. 684-721.

---, 1985, *Principles of Structural Geology,* Prentice-Hall, Englewood Cliffs, NJ.

---, 1991, Analysis of Scale, v. 75, No 7, p. 1161-1177.

Suppe, J., G. T. Chou, and S. C. Hook, 1992, *Rates of folding and faulting determined from growth strata, Thrust Tectonics,* K. R. McKlay (ed), Unwin Hyaman Pub., p. 105-121.

Suppe, J., and D. A. Medwedeff, 1984, *Fault-propagation folding, Abstracts with Programs:* GSA Bull., v. 16, p. 670.

---, 1990, *Geometry and kinematics of fault-propagation folding,* Eclogae Geol. Helv., v. 83/3, p. 409-454.

Tearpock, D.J., and R. E. Bischke, 1990, *Mapping throw in place of vertical separation: a costly subsurface mapping misconception,* Oil and Gas Journal, July 16, v. 88, No. 29, p. 74-78.

---, 1991, *Applied Subsurface Geological Mapping,* Prentice-Hall, Englewood Cliffs, NJ.

Tearpock, D. J., and J. Harris, 1987, *Subsurface Geological Mapping Techniques - A Training Manual,* Tenneco Oil Co., Houston, TX.

Thorsen, C. E., 1963, *Age of growth faulting in southeast Louisiana:* Trans. - GCAGS, v. 13, p. 103-110.

Tucker, P. M., and H. J. Yorston, 1973, *Pitfalls in Seismic Interpretation,* Soc. of Exploration Geophysicists, Monograph No. 2.

Vail, P. R. and W. Wornardt Jr., 1991, *An integrated approach to exploration and development in the 90's: Well log-seismic sequence stratigraphic analysis,* Trans. GCAGS, v. 31, p. 630-650.

Vogler, H. A., and B. A. Robinson, 1987, *Exploration for deep geopressure gas: Corsair Trend, offshore Texas:* AAPG Bull., v. 37, No. 1, p. 158-162.

Wadsworth, A. H., Jr., 1953a, *Percentage of thinning chart-new techniques in subsurface geology:* AAPG Bull., v. 37, No. 1, p. 158-162.

282

Webber, K. J., and E. Daukora, 1976, *Petroleum geology of the Niger Delta:* Ninth World Petroleum Congress, v. 2, p. 209-221.

Wharton, J. B., Jr., 1948, *Isopachous maps of sand reservoirs:* AAPG Bull., v. 32, No. 7, p. 1331-1339.

Wolff, C. J. M., and J. P. Dewardt, 1981, *Borehole position uncertainty-analysis of measuring methods and deviation of systematic error model,* Jour. of Petroleum Technology (Dec., 1981), p. 2339-2350.

Wright, T. L., 1991, *Structural geology and tectonic evolution of the Los Angeles Basin, California, in Active Basin Margins,* AAPG Mem. 52, K. T. Biddle, ed. p. 35-134.

Xiao, H., and J. Suppe, 1989, *Role of Compaction in Listric Shape of Growth Normal Faults:* AAPG Bull., v. 73, No. 6, p. 777-786.

---, 1992, *The origin of rollover,* AAPG Bull., v. 76, p. 509-529.

Yeats, R. S., and J. C. Taylor, 1990, *Saticoy Oil Field - U.S.A., Ventura Basin, California, in Beaumont,* E. A. and Foster, N. H. eds, AAPG Treatise of Petroleum Geology, p. 199-219.

INDEX

A

Anomalous velocity layers, 78-79
Antilistric fault bends, 178
Area conservation principle, 102
Axial Surfaces, 117-120, 126, 131, 134
    locating fault position, 183-186

B

Balanced cross sections, 4
Basinward closure, 195-196
Bed dip geometry, 84
Bed length consistency, 103
Bed thickness, 103
Bow and Arrow Rule, 140
Box folds, 136
Brittle deformation, 103

C

Chute, 169
Coherent noise, 78-79
Common fold patterns, 188
    rolled up beds, 189-191
    half-graben monoclinal rollover structure, 192-195
    fanning of bed dips, 195
Compressional settings:
    types of structures, 138
Conserving area, 102

Continuous reflectors, 89
Contour compatibility, 16-19, 39
    across faults, 39
    closely spaced horizons, 16
Contour lines:
    odd number of, 76-77
Contouring:
    equal-spaced contouring, 12-16, 36
    interpretive contouring, 14
Correlation and mis-tie techniques, 87-94
    continuous reflector, 89
    cross line ties, 87
    tying wells, 88-89
Correlations:
    importance of accuracy of, 4
Coulomb shear, 34,97
Cross contour, 41, 43, 54, 56
Cross sections:
    balanced, 4, 157
    fish-hook faults, 159-160
    multi-play, 161
    passive roof duplex, 159
    types of,
        illustration aid, 27, 28
        study aid, 16
    vertical exaggeration, 121

284

D

Dêcollments, 130, 161-162
Δd/d technique, 212-220
Dip domain analysis, 126-129
Directional Wells, 224-236
    Net pay isochore map, 232-236
    Net sand map, 232-236
    Production faults, 229-232
    Vertical separation, 224-229
    Water level changes, 229-232
Documentation, 2
Downward dying growth fault, 220
Dry hole analysis, 6-11
    reasons for, 6-7
        geologic, 7-11
Duplex structures, 116, 141-146
    Dip spectral analysis, 143-146
    Second order duplex, 145-146
    Third order duplex, 143-145

E

Equal-spaced contouring, 12-16, 36
    map constructed using, 13
    Tee-Pee structure, 14
Expansion indices, 212-220
    Thorsen Expansion Index, 213
Extensional settings:
    types of structures, 138

F

Fault and structure map integration, 4, 40-46, 46-54
Fault cut data, 40, 46
Fault gap, 58
    definition, 67
    vs fault heave, 67-70
Fault heave:
    definition, 67
    vs. fault gap, 67-70
Fault propagation fold, 107, 120
    Dip analysis, 162-167

Fault surface:
    analysis of, 34
    related to fold shape, 113
Fault traces, 40, 103
    Rule of 45, 40, 65-67
Faulted structure, 33
Faults:
    additive property of, 57-60
    fish-hook, 159-160
    growth, 10, 35-36, 39, 96-97, 99
    listric growth normal, 34, 114, 168-169
    normal, 186
    reverse, 37-38, 59
    saucer, 74-76
    screw, 64, 70-74
    sealing nature of, 10
    structure map, implied analysis of, 60-64
    thrust, 37, 39, 111, 125, 131
    wrench, 39, 136
Fold limb bisecting rule, 117
Fold Shape:
    related to fault surface, 113

G

Growth:
    expanded reflections, 211-212
    extensional structures:
    antithetic fault termination, 203-204
    axial surfaces, 204-209
      slope of, 209-211
    compressional structures:
      growth axial surfaces, 152-156
      slow sedimentation, 156
    faults, 35-36, 96-97, 99
      Δd/d method, 10, 212-220
      downward dying, 220
      Expansion Index, 10, 212-220

H

High-Low mapping problem, 28-32
Hydrocarbon traps:
    conditions, 7
    not present,
      reasons why, 9
    uneconomic, 10

I

Illustration aid:
    structure map accuracy, 52-54
Incorrect time migration, 83
Inner limit of water, 238
Interpretive contouring, 2, 5
    map constructed using, 14
Inversion structures, 137
Isochore map, 234, 237-247
    edge water, 237
    net gas, 239
    net pay, 232-236, 241-245
      Wharton Method, 237-241
    versus isopach, 246-247
Isopach map, 234
    versus isochore, 246-247

K

Kink method, 107

L

Listric growth normal faults, 34
    determining rollover shape, 168-169

M

Missing section:
    definition, 46
    and vertical separation, 46-54
Multiple horizon mapping, 4

N

Near surface velocity anomalies, 78-79
Net gas isochore map, 239
Net pay isochore map, 232-236, 241-245
    Wharton method, 237-241
Net sand map, 232-236
Normal faults, 34, 40, 46-54
    growth, 35-36, 39, 96-97,99
    relationship with prospect   shape and position, 186-188
      internal deformation, 188

O

Out of plane reflectors, 96

P

Philosophical doctrine, 1-5
Porosity bases, 21-22
Porosity top mapping, 20
Prospect shape and position:
    relationship with normal faults, 186-188
      internal deformation, 188

R

Reservoir section,
    absent or tight, 8
Restored tops, 23
    diapiric structures, 25
    methods of using to check map accuracy, 25
Retrodeformation, 109
Reverse faults, 38, 40, 59
    structure map integration, 54-57
Rule of 45, 4, 65-67

S

Saucer faults, 74-76

Screw fault map, 72, 73
Screw faults, 70-74
   definition, 70
Seismic data, 78-101
   Bed dip geometry, 84
   Quality and validity, 78-87
      Coherent noise, 78-79
   Velocity problems, 79-87
      anomalous velocity
      layers, 79
      incorrect time
      migration, 83
      near surface velocity
      anomalies, 79
      velocity gradients, 81
Side-swipe reflectors, 84, 96
Strike lines, 175-176
Strike ramp, 222
Structural compatibility, 38
   across faults, 39
   reasons for incompatibility, 39-40
Structure maps:
   accuracy of, 49-54
   faulted, 33-77
   faults, additive property of, 57-60
   general, 12-32
   High-Low problem, 28
   implied fault analysis, 60-64
   reverse fault integration, 54-57
   unrealistic, 13
Subsurface mapping:
   philosophical doctrine of, 2-5

T

Tee-Pee structures, 14, 28
Ten-minute map, 8
Thin skinned tectonics, 111
Three-dimensional:
   structural geometry, 95-100
   validity, 3
Throw, 46-54, 57
   mathematical relationship of
   vertical separation to, 50
   measurement of, 51-52

Thrust Faults, 37, 39, 111, 125, 131
   ramp faults, 38-39
   Related to compressional structures,
   111, 125, 131
Transfer zones, 139
True stratigraphic thickness, 234, 246
True vertical depth thickness, 228
True vertical thickness, 224-229, 232,
246

V

Velocity data:
   problems with, 79-87
      anomalous velocity layers,
      79
      incorrect time migration, 83
         multiples, 83
         undermigration, 83
         overmigration, 83
      near surface velocity
      anomalies, 79
      velocity gradients, 81
Velocity gradients, 81-82
Vertical Exaggeration, cross sections, 121
Vertical Separation, 46-54, 57, 58
   definition, 47
   directional wells, 224-229
   mathematical relationship of throw to,
   50
   seismic lines, 89
Volume Conservation Principle, 187

W

Wharton Method, 237-241
Wrench:
   faults, 39, 136
   types of structures, 138